ABSTRACT ALGEBRA

AND SOLUTION BY RADICALS

JOHN E. MAXFIELD

Professor of Mathematics
Kansas State University

MARGARET W. MAXFIELD

W. B. SAUNDERS COMPANY   PHILADELPHIA · LONDON · TORONTO · 1971

W. B. Saunders Company: West Washington Square
Philadelphia, Pa.   19105

12 Dyott Street
London,   WC1A   1DB

1835 Yonge Street
Toronto 7, Ontario

Abstract Algebra—and solutions by radicals   SBN 0-7216-6187-4

Print No.:   9   8   7   6   5   4   3   2   1

DEDICATION

This book is lovingly and respectfully dedicated to our parents, C. G. and Lucile Maxfield and Frederick V. and Irma W. Waugh, and to our other fine teachers, including a neurotic magazine-straightener named Thompson, I think, who kept drumming it into his seventh grade science class that precise statement was the key to science, to Forrest H. Keck, to Mr. Dresia, and to the memory of Mr. Herner, to the memory of Professor C. H. Yeaton, who gave the most intriguing mathematics tests ever devised, to the memory of Professor C. C. MacDuffee, who handed the chalk to his students, and to Professor Ivan Niven, who suffered not one but two dissertations at our hands.

In the intensive editorial development carried out on the book we received valuable help from Professors Arlo W. Schurle of Indiana University, Allen B. Altman of the University of California at San Diego, Herbert J. Nichol of Drexel University, Guy T. Hogan of SUNY State College at Oneonta, and Brother Hugh N. Albright of LaSalle College, each of whom critically reviewed the entire manuscript. Their detailed and percipient comments resulted in many improvements to the final manuscript. We are grateful to George Fleming, our editor at W. B. Saunders Company, who encouraged and organized the project from the first tentative ideas.

CONTENTS

HOW TO USE THIS BOOK

The main text of this book is unusually full, especially for a mathematics book, which is often so hard to read that an ordinary literate adult is reduced to having it read to him by the instructor. We hope to encourage you to read for yourself.

To keep you from missing the basic organization in the admittedly wordy text, we have set off the definitions and theorems with a bracket. These form the skeleton of the book. When reviewing, simply go over this skeleton, re-reading the explanations in the text where necessary.

You will notice that some sections of the text are shaded, as, for example, some of the material on isomorphisms surrounding Definition 2–15. These are the sections we consider part of your general education in this Age of Automation. The whole book can bring you to an appreciation of abstraction and of arbitrary postulational systems, ideas which are central in automation. The ideas explained in the shaded sections are especially important and characteristic of the kind of mathematics we are covering. Most of them lend themselves to ordinary discussion among laymen without any lengthy introduction or special jargon. Their flavor will stay with you even if you forget specific theorems. That flavor is especially useful to a teacher, for although he may not teach such advanced material, he will want to present elementary material using the most modern approach in a way consistent with what the student will learn if he goes farther.

Some exercises are labeled "Classroom Exercise." Attentive participation while the whole class works these exercises together will form a bridge between passive listening and active homework.

The "Historical Intermission" can be used in several different ways. In our own pilot classes we have encouraged students to relax with it on their own, discussing it if they seem to want to.

The Appendices can be used to further your interest along several different lines barely suggested in the text. They offer an opportunity for outside work to bolster a sagging grade. Some contain proofs postponed from various chapters so as not to impede the continuity of thought on a first reading.

THE NEGATIVE OBJECT
OF THIS BOOK

Perhaps you remember a way of solving an equation like $3x^2 - 13x - 10 = 0$ by the "quadratic formula" for the roots r_1 and r_2 of the equation $ax^2 + bx + c = 0$:

$$r_1, r_2 = \frac{-b \pm \sqrt{b^2 - 4ac}}{2a}.$$

If the coefficients 3, -13, and -10 of the equation are used respectively as a, b, and c, the formula gives the two roots

$$r_1, r_2 = \frac{-(-13) \pm \sqrt{(-13)^2 - 4(3)(-10)}}{2(3)} = 5, -2/3.$$

Then we say we have "solved" the equation by finding all its "roots," by which we mean all the numbers that make the equation a correct statement about numbers when used as x:

$$3(5)^2 - 13(5) - 10 = 0 \quad \text{and} \quad 3(-\tfrac{2}{3})^2 - 13(-\tfrac{2}{3}) - 10 = 0.$$

Later as we develop precise vocabulary we shall define our terms, but not for the purpose of introducing more formulas.

In this book we focus on a *negative* result, Abel's Theorem, that there can be no formula for solving polynomial equations of degree greater than four; that is, no formula that will work in all cases and that involves a finite number of additions, subtractions, multiplications, divisions, and extractions of roots. Also, we study in Appendix M the related negative result that there is no ruler-and-compass construction for trisecting an arbitrary angle.

We do not say, "No formula has been found . . ." or "You cannot solve . . ." or "Students with an I.Q. less than 105 cannot. . . ." We say, "There can be no formula. . . ." For this reason mathematicians have to supply just as rigorous a proof for such a negative result as for a positive one. They must satisfy themselves not just that they have failed to turn up such a formula but that no such formula could possibly exist.

One advantage to studying this negative result is that it clarifies the arbitrary "game" nature of mathematical deduction. We are not going to prove that every equation of degree five is completely unassailable. Abel's Theorem proves only that in general the roots cannot be found *according to the rules established*.

As an analogy, take the old game of making words from the letters of another word. Suppose you are given. the word A L G E B R A. Can you make the following words from it? GEAR, LEG, BEAR, BALL? What about the last one? May we use "L" twice when it appears only once in the given word "ALGEBRA?" That depends on the rules of the game, and we must at this point decide with more precision just what those rules are. Notice that all sorts of other rules might be imposed, depending on the way we want the game to develop. For instance, we might insist that letters be used in the same order in which they appear in the given word, which would cut down greatly on the number of possible (and here we really mean "allowable") solutions. The rules in mathematics may be considered arbitrary in this same sense, as they are purposely altered to produce different mathematical results. (For instance, altering rules in geometry to provide for not one, but at least two distinct parallels to a given line through a given point not on the line produces a geometry different from classical Euclidean geometry.) Now if we ask "Can you spell 'GEOMETRY'?" your answer depends on context. Presumably, your spelling ability is up to the task, but are you still spelling according to the rules of the "ALGEBRA" game? If you are, the answer is negative.

Abel's Theorem is this kind of negative result, to be understood in the context of the rules of a certain mathematical "game," or deductive system. In the course of proving Abel's Theorem we explore the context of rules surrounding the problem of solution by radicals and in so doing arrive at a general viewpoint from which elementary algebra can be seen in perspective as a coherent organized whole.

An important strategy, especially in proving a negative result, is a "contrary to fact" or *"reductio ad absurdum"* attack. In such a proof we assume the contrary to what we want to prove and on the basis of that assumption deduce things we know cannot possibly be valid. A mother might use such a strategy in proving Junior had not washed his hands before coming to dinner: Suppose he had washed them. Then the sink and surrounding area would be spattered with muddy drops and the towel would be dirty and crumpled on the floor. But inspection shows a bathroom of unblemished tidiness. Therefore the assumption that he had washed has led to deductions of obvious absurdity. Watch for many examples of *reductio ad absurdum* proofs here, including the whole chain of theorems leading to Abel's Theorem.

Although you will find yourself reviewing elementary algebra as you read this book, the review is only incidental. In fact, the book does not begin with a review, but starts directly with the very general and important definitions of binary operations and groups. In the first part of the book you will learn about groups, rings, fields, and mappings, in enough detail and with enough examples to make you comfortable with the definitions. Then in the latter part of the book you will see how the separate topics interact in the study of Abel's Theorem. Here the object is no longer complete mastery to the point of comfort with the material, but perception of the possibilities inherent in abstract algebra. Most readers will not become expert at applying the Galois theory of Chapter 8, but they will retain the impression of structure in algebra that will color all their later contact with the subject.

GROUPS

Definition 1–1. A **binary operation** θ on a set S is a mapping of $S \times S$ into S. This means that for every pair a, b of elements of S there is an answer to the question $a \,\theta\, b = ?$ and that answer is some definite (single) element c of S.

  i. S is then **closed** under θ. **Closure** is the word we use to say that if the problem $a \,\theta\, b = ?$ is made up of elements a and b in S, then the answer is also in S.

Mappings will be defined in Definition 2–12, and undefined terms like "set" will be discussed on page 21 , but first we build up some experience with special cases, to sharpen intuition.

The operation $+$ of addition applied to the counting numbers 1, 2, 3, . . . provides an example of a binary operation θ, since to each ordered pair, such as (23, 48), there corresponds a definite sum (23 + 48 = 71).

The essential closure property guarantees that the binary operation will continue to yield answers in the original set. One advantage of this is that we can apply the operation several times in sequence, as in the successive additions $[(18 + 3) + 7] + 14$. Closure is easily explained in connection with a system that does not have it. Take, for instance, all fractions that are written in lowest terms with denominators of either 2 or 3, such as $\frac{1}{2}$, $\frac{2}{3}$, $\frac{22}{3}$, $\frac{39}{2}$, and so forth. Is this system closed under the usual addition of fractions? To find out, try $+$ in place of θ in the definition of closure, and test whether $a + b$ is always in the

1

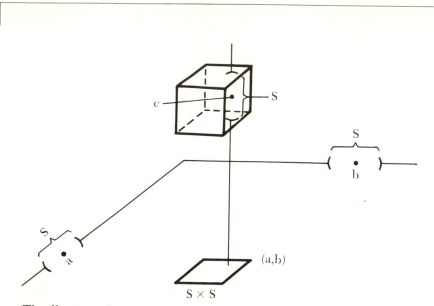

The direct product set $S \times S$ is made up of ordered pairs (a, b) of elements a and b of the set S. A binary operation θ yields for each (a, b) in $S \times S$ a single element c of S.

FIGURE 1–1.

system of fractions with denominators 2 or 3 when a and b are. What if $a = \frac{1}{2}$ and $b = \frac{1}{3}$? We have $a + b = \frac{1}{2} + \frac{1}{3} = \frac{3}{6} + \frac{2}{6} = \frac{5}{6}$, which cannot be reduced so as to have a denominator of 2 or 3. This means that the system is not closed under $+$, and that $+$ is not an operation on the system, though it is an operation on the system of all fractions.

Classroom Exercise 1–1. Are the even positive integers closed under $+$? under $\times$?

Classroom Exercise 1–2. For rectangles there corresponds to each length measurement-width measurement pair an area measurement. Still, computing the areas is not, strictly speaking, a binary operation. Why?

Exercise 1–3. Are the odd positive integers closed under $+$? under $\times$?

Exercise 1–4. Show that the system of fractions having denominators of 1, 2, or 4 when written in lowest terms is closed under $+$.

Exercise 1–5. Show that the system of all fractions expressible in lowest terms with denominators of 18 or any divisor of 18 is closed under $+$, and formulate a general theorem about any integer d and all its divisors. Indicate how to generalize the proof for $d = 18$ to a proof for any integer d.

Exercise 1–6. Are the counting numbers 1, 2, 3, . . . closed under subtraction?

Definition 1–2. A **group** G is a set of elements $a, b, c, \ldots$ with a binary operation $*$ satisfying three postulates:

    i. $*$ is **associative** in G, *i.e.*, for every triple a, b, c in G, $(a * b) * c = a * (b * c)$. The answer does not depend on the order in which the two-stage operation is performed.

    ii. G has an **identity** e with respect to $*$, *i.e.*, G has an element, which may be called e, with the property that combining it with any element a leaves a unchanged; $a * e = e * a = a$, for every a in G.

    iii. Each element in G has an **inverse** with respect to $*$, *i.e.*, for each a in G there is an element called a^{-1} (read "a-inverse") for which $a * a^{-1} = a^{-1} * a = e$, an identity element of G.

    We assure nontriviality by assuming that G has at least the one element e.

We introduce some other closely related definitions and notations before enlarging on all these new concepts together.

Definition 1–3. The number $|G|$ denotes the **order** of G, that is, the number of elements in the group G. A **finite** group of order $|G|$ has elements that can be put into one-to-one correspondence with the counting numbers $1, 2, 3, \ldots, |G|$.

Definition 1–4. We write $a \in \mathbf{G}$ for "a is an element of G," "a is a member of G," or "a is in G."

Discussion of the Group Properties

The associative property guarantees that if we combine three elements, keeping them in the same order, we obtain the same answer no matter which of the two pairs we elect to combine first. Our experience with nonassociative systems is limited, so we have to reach a bit for examples to point up associativity, but one nonassociative feature in mathematics is "exponentiation," or raising-to-powers. We will say that $a \theta b$ in this case means "a raised to the power b" or a^b. In this case $(a \theta b) \theta c$ does not necessarily equal $a \theta (b \theta c)$, for consider

$$(2^3)^2 = 8^2 = 64 \quad \text{and} \quad 2^{(3^2)} = 2^9 = 512.$$

Classroom Exercise 1–7. Compute $[(1.2)(0.3)](2.7)$ and $(1.2)[(0.3)(2.7)]$, rounding to the nearest tenth after each multiplication. Conclude that multiplication with rounding is not associative.

To get outside the realm of mathematics for a moment, let θ denote "meets" applied to "persons." It may make quite a difference whether Tom meets Mary and later that couple meets Dick or whether Tom meets the couple Mary-and-Dick.

$$(T \, \theta \, M) \, \theta \, D \neq T \, \theta \, (M \, \theta \, D).$$

Classroom Exercise 1–8. Think of a system that fails to have the associative property. It need not be mathematical.

The identity depends on the operation $*$, as it is an element e that leaves every element unchanged when combined with it by the binary operation. Consider the integers, including the negative integers and zero. Under the operation $+$ the identity is 0, because $a + 0 = 0 + a = a$ for every integer a. Under the operation $\times$, though, the identity e is 1, because $a \times 1 = 1 \times a = a$ for every integer a.

Exercise 1–9. Suppose three dancers in a line occupy positions L(eft), M(iddle), and R(ight), and that the elements of our set are the six different ways they can change their positions; for instance, two elements of the set are

$$a \begin{cases} \text{Whatever dancer is in position } L \text{ can move to } M. \\ \text{The dancer in position } M \text{ can move to } R. \\ R \text{ can move to } L. \end{cases}$$

and

$$b \begin{cases} \text{The dancer in position } L \text{ can move to } M. \\ M \text{ can move to } L. \\ \text{The dancer in position } R \text{ can stay in } R. \end{cases}$$

The operation $\circ$ combining two of these changes will be the *composition* of the two changes; that is, the net result of one change followed by the other. We shall make first the position change named to the right of the operation sign $\circ$, so that $a \circ b$ is the change that results if we make change b followed by change a. (See Fig. 1–2.) What "change" acts as an identity for such a system?

The inverse is defined in terms of an identity e as an element a^{-1} that combines under $\circ$ with a given element a to produce e. Thus, an identity works for all the elements of the group, but each element has its own inverse. The inverse of an integer a with respect to $+$ is its negative, $-a$; in fact, that is the way we introduce negative integers. What is the inverse of a fraction with respect to $+$? What is the inverse of a positive integer with respect to $\times$? Is it an integer? Consider a system made up of the positive integers and their inverses with respect to $\times$ (also called their "reciprocals"). Is this system closed with respect to $+$?

a   ∘   b   =   a ∘ b

L	M	R

In change b dancer ✖ moves from L to M; then in change a ✖ moves from M to R. Then the net change for dancer ✖ in the composition $a \circ b$ is a move from L to R. In $a \circ b$ dancer | moves from M to M, and dancer Σ moves from R to L.

FIGURE 1–2.

Exercise 1–10. Return to Exercise 1–9. What is the inverse change of the change a? What is the inverse of b?

Exercise 1–11. Show whether the system $\{1, r, s\}$, θ is a group if the elements combine under θ as shown in the following table. (In order to find $r \theta s$, for instance, look in the horizontal row labeled "r" and in the vertical column labeled "s"; $r \theta s = r$.)

θ	1	r	s
1	r	s	1
r	s	1	r
s	1	s	r

Further Properties of Groups

The definition of a group gives us three requirements that a set with operation must satisfy to qualify as a group. There are other properties that every group has that we could have listed, but we prefer to derive or prove them from the three we have given. The next two theorems are examples of further properties of groups that we have not adopted explicitly as postulates but which are implied by the three group postulates we have adopted.

Theorem 1–1. The identity of a group is unique.

PROOF: If a group had more than one identity element, we could call two of them e and e'. Form the combination $e * e'$. Because e is an identity, the combination would equal e'; that is, $e * e' = e'$. But because e' is an identity, the combination would also equal e; that is, $e * e' = e$. Since $*$ is an operation, there is exactly one element $e * e'$, and e and e' must be two different designations for it. Then $e = e'$. ∎†

Let us comment on this ingenious little proof. As we said, we could have assumed the result explicitly along with the three properties assumed for any group. In a sense we can say we assumed the result implicitly among the three properties, because we proved it by using the definition of an identity in property *ii*.

In an effort to further mathematical creativity, let us see how we might have gone about constructing a proof for Theorem 1–1. Suppose someone gave us the elements and operation of a large complicated system and claimed that two different elements, e and e', were identities. Probably one of the first experiments that might occur to us would be checking the multiplications to make sure both e and e' really behaved like identities, that is, that they both left each element unchanged when combined with it by the operation. If the first few elements were unchanged when combined with e, we might think of the fact that one particular element, e', is of particular interest to check. Similarly, it should be of special interest to combine e' with e.

Theorem 1–2. In a group the inverse of each element a is unique.

PROOF: Suppose there is some element a having two inverses. Give them different symbols, a^{-1} and a'. Refer to group property *i* to write $a^{-1} * a * a'$ in two equal ways. Use property *iii* to replace each combined pair by the (unique, from Theorem 1–1) identity e. Then use property *ii* to show that the "two" inverses are really equal. ∎

Exercise 1–12. Fill in the details of the proof of Theorem 1–2.

Are there any redundancies (unnecessary repetitions) in Definition 1–2? We did not include uniqueness of the identity nor uniqueness of inverses as assumptions because we could deduce them from the things we did assume, as in Theorems 1–1 and 1–2. Did we assume any properties that could have been deduced as theorems from the rest? Yes, we did. To discuss them it will help to have another definition.

---

† This symbol, ∎, means we have finished the proof.

Definition 1–5. A binary operation θ on a set S is said to be **commutative** if for every pair a, b $a \theta b = b \theta a$. We also say then that S is· commutative under θ, that a and b commute under, or with respect to, θ, and that θ has the property of commutativity.

Notice that the associative property given as the first group requirement does not involve commutativity of the elements but involves only a change in the order of combining three elements, the three remaining in the same order.

Definition 1–6. A group in which the operation is commutative is called· a **commutative group** or an **abelian group.** The word "abelian" honors Abel, a biography of whom is included in this book.

Notice that in Definition 1–2 we did not require the group operation to be commutative, but we did assume in *ii* that the identity e commutes with every element and in *iii* that each element commutes with its inverse. These two special cases of commutativity need not have been assumed, as they could have been deduced. That is, we could have assumed a right (or a left) identity for the group and for one of the right identities a right inverse for each element. From these assumptions and property *i* we could have deduced that any identity commutes with each element of the group, that the identity is unique, and that the inverse for each element is unique and commutes with the element. (This is done in Richard A. Dean's *Elements of Abstract Algebra*, Wiley, 1966, pp. 30–31.)

If the weaker assumptions would have sufficed, *should* we have used them instead of the stronger ones? Is it *wrong* to include some redundancy in our definition? These words "should" and "wrong" apply to moral questions we do not treat mathematically. If there is any basis for our decision here it is esthetic, rather than moral. For some reason it pleases most of us to minimize the restrictions we impose and still have a definition that exactly fits the objects we want to define. In this case we choose the stronger restrictions, however, because we want to work directly toward solution by radicals without taking time to deduce the stronger restrictions from the weaker ones.

This is a good place to emphasize the arbitrary character of mathematical premises. It is up to us where we start. We could have assumed the uniqueness properties proved in Theorems 1–1 and 1–2 along with the group postulates, but we chose to include the two "elegant" short proofs—another case of esthetic motivation in mathematics. In this book we shall assume many of the properties of the arithmetic of integers, rational numbers, and real numbers, reviewing what they are but not proving that they can be deduced from postulates. Each of these sets of numbers, however, can be explored on the basis of a small set of postulates and definitions (see, for instance, A. H. Lightstone's *Symbolic Logic and the Real Number System*, Harper and Row, 1965). It is as if we planned to test part of a long chain to make sure each link tested connects firmly to the one before it. We are most interested in a certain section of the chain, so we merely suppose it has passed all tests up to that section.

Mathematical premises are so arbitrary that they may not even be realistic. That is, they may not appear to be "so" or "true" in reality. Still, in mathematics we can go right on deducing theorems from these premises and are happy enough if some of the results are "interesting." The theorems, then, we call "valid," not necessarily "true." "Valid" and "true" are as different as "acquitted" and "innocent." Abraham Lincoln once brought his Cabinet back to earth by asking them, "If we call a cow's tail a leg, how many legs does the cow have?" When they answered, "Five," he corrected them, saying that a cow would have four legs no matter what you called her tail. This story points up beautifully the difference between mathematics and science, for in science the cow would have four legs, but in mathematics she would have five if her tail was assumed to be a leg.

How can an obviously false assumption ever be useful? Consider the many practical uses of false assumptions like "The Earth is a sphere," "The Earth is an oblate spheroid," "Students who score highest on exams know the most," "This experiment takes place in a vacuum," "The coefficient of friction is zero," "The growth curve is exponential," and so forth.

Classroom Exercise 1–13. Take the system of integers, *i.e.*, the counting numbers, their negatives, and 0, with a combining rule θ defined by $a \theta b = ab - b$, for any integers a and b. Is the system commutative? closed under θ? associative? Is there an identity element (two-sided)? Then, what can you conclude about the inverse property?

Exercise 1–14. Take the system $\{0\}$, θ with multiplication table

$$
\begin{array}{c|c}
\theta & 0 \\
\hline
0 & 0
\end{array}
$$

(that is, $0 \theta 0 = 0$). Is it a group? Is it a commutative group? Does it represent the only possible 1-element group, disregarding notation?

Exercise 1–15. Let θ select the maximum of positive integers a and b, so that $a \theta b = a$ if $a \geq b$ and $a \theta b = b$ if $b > a$. Is θ a binary operation on the counting numbers 1, 2, 3, ...? Is the system associative? Commutative? Has it the identity property? the inverse property?

Examples of Groups

1. The Klein four-group (named for Felix Klein [1849–1925], geometer and master teacher). We can define multiplication of elements of the set $\{1, a, b, c\}$ by the following table:

$\cdot$	1	a	b	c
1	1	a	b	c
a	a	1	c	b
b	b	c	1	a
c	c	b	a	1

The entry in the ith (horizontal) row and jth (vertical) column in the body of the table gives the product of the element in the ith row heading and the element in the jth column heading, in that order. For example, $a \cdot c$ has the value b, $b \cdot a$ the value c.

We can prove that the set $\{1, a, b, c\}$ with multiplication as specified by the table is a group. To verify this, we must first verify that the multiplication defined by the table constitutes a binary operation on the set, according to Definition 1–1. It is indeed defined on $S \times S$, for the body of the table has an entry for every pair of headings s, t. Each product $s \cdot t$ is uniquely defined, for there is exactly one product entry for each position in the body of the table. All entries in the body of the table are elements of S, that is, are 1, a, b, or c, so that S is closed under multiplication.

Next we have to verify that all requirements of Definition 1–2 are met. The hardest requirement to check is that of associativity, which theoretically calls for trying all triples that can be formed from the four elements. We can shorten the work with some analysis of the situation, but first we check the easier group requirements: The element 1 plays the role of the identity of the group, for it does not change any element upon multiplication from left o right. How does the multiplication table enable us to find the inverse of each element? The inverse of an element s is the element s^{-1} that multiplies it to give the product 1. To find s^{-1} we can follow along the row headed "s" until we find a product 1, then find the inverse as the column heading above the 1. To be sure that each element does have an inverse, we need only verify that there is a 1 in each row. For instance, to find b^{-1}, we read along the third, or b, row until we find 1 in the third column, whose heading is b; so $b^{-1} = b$.

Now to cut down on the work of demonstrating associativity, notice that this group has the commutative property of Definition 1–5; that is, we could swap rows and columns without changing the multiplication table, because it is symmetric about the 1's in the main diagonal. By the commutative property, we see that for a single element s, $(s \cdot s) \cdot s = s \cdot (s \cdot s)$. Therefore, the associative property holds for any triple made up of just one element repeated three times. Also, from commutativity, we conclude that if stu is an associative triple, then so is uts, because $s(tu) = (st)u$ can be changed by commuting elements into $(ut)s = u(ts)$. Next, it is easy to check triples one of whose elements is the identity 1. For instance, $1(st) = st$, by the property of the identity element, and $(1s)t = st$, so they equal each other, since the operation gives exactly one result. We can treat any triple of the form $s(ts)$ by commutativity: $s(ts) = s(st) = (st)s$. The remaining cases can be checked one by one.

$a(bc) \overset{?}{=} (ab)c$. We use the multiplication table to reduce the left member, $a(bc)$, to $a(a) = 1$. The right member, $(ab)c = (c)c = 1$, so they are equal.

Classroom Exercise 1–16. Similarly, verify that

$$a(ab) = (aa)b \quad a(cb) = (ac)b$$

$$a(ac) = (aa)c \quad b(ba) = (bb)a$$

$$b(ac) = (ba)c \quad c(ca) = (cc)a$$

$$b(bc) = (bb)c \quad c(cb) = (cc)b$$

We have proved that the set-and-operation is a group, because it satisfies the requirements of Definition 1–2. In fact, it is an abelian group, since it satisfies the additional requirement of Definition 1–5.

Exercise 1–17. Take the system $\{1, x, y, z\}$, θ with multiplication table:

θ	1	x	y	z
1	1	x	y	z
x	x	y	z	1
y	y	z	1	x
z	z	1	x	y

Is it a group? Is it a commutative group? Is it just the same as the Klein four-group except for different letters? (Look at $x \, \theta \, x$, for instance.)

Application.   Has the four-group any use, any application besides illustrating the group properties? Here is a simple physical realization:

red switch   green switch

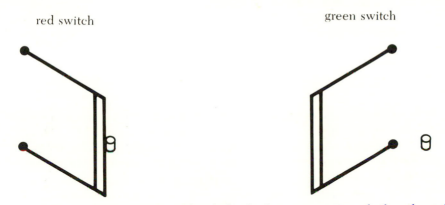

Let *a* stand for switching the red switch. It does not matter whether the red
switch is open or closed initially; *a* stands for changing its setting from what it
was. Let *b* stand for switching the green switch.

Now we can figure out how to assign *c* and 1 to make the four different
switch operations into a realization of the four-group. As our binary operation
we use *composition*; that is, we define *a* ∘ *b* to mean "do *b*, then do *a*," that is,
switch the green switch, then the red one. Since the product *a* · *b* from the
multiplication table is *c*, we let *c* stand for switching both switches. Since *a* · *a*
is 1, let 1 stand for leaving both switches unchanged, producing a result equivalent to any even number of changes in each switch. A little experimentation
will convince you that the two switches with the four ways of changing their
settings and the combining operation of composition give a laboratory equivalent to the four-group.

It is easy to study the limited potentialities of the two switches without the
"mathematical model" of the four-group, but we can use the example to suggest
far more complicated circuitry, as in computing machines. In a complicated
laboratory set-up, the introduction of an ingenious abstract model, such as a
group, may make analysis easier.

We can make two important observations about the switch example,
analogous to two we made about the four-group. First, the changing of the
switches is commutative under composition, for it makes no difference to the
final setting of the switches which of the two changes we make first. Second,
the changes are associative under composition, for *s* ∘ (*t* ∘ *u*) means "perform
whatever change is equivalent to change *u* followed by change *t*; then perform
change *s*," while (*s* ∘ *t*) ∘ *u* means "perform change *u*; then perform the change
equivalent to change *t* followed by change *s*." We see that both *s* ∘ (*t* ∘ *u*) and
(*s* ∘ *t*) ∘ *u* produce the same result as change *u* followed by change *t* followed by
change *s*.

The proof of associativity in this example can be used to suggest a general
theorem that will often help us avoid a tedious proof of associativity "by
exhaustion" (trying all possible combinations).

Theorem 1–3. The mappings from a set into itself are associative under composition.

PROOF: First, we enlarge a bit on the economical language of the theorem.   Let s, t, and u be any three mappings of a set A into itself; that is, s maps each set member $x \in A$ into a certain member, $x^s \in A$, t maps any $y \in A$ into some $y^t \in A$, and so on.   As an example, suppose that A is a set of 5 people, each of whom holds one of 5 positions in an organization, such as offices in a municipal government or administrative positions in business management.   At intervals the 5 positions are shuffled among the 5 people, and a ceremony is held at which each person installs the person who replaces him in the position he is relinquishing.   Then each installation constitutes a map from A to A, where the image of a person is the one he installs.   For instance, at the installation

$$s = \begin{pmatrix} C & L & N & E & G \\ N & C & G & L & E \end{pmatrix}$$

Charles installs Niels as his successor in office, Ludwig installs Charles, Niels installs George, Evariste installs Ludwig, and George installs Evariste.

Let the binary operation on any two mappings s, t be composition of the mappings; that is, $s \circ t$ means the mapping t followed by the mapping s applied to the result.   In the example, suppose an installation

$$t = \begin{pmatrix} C & L & N & E & G \\ E & L & N & G & C \end{pmatrix}$$

is followed by the installation s.   The composition $s \circ t$ produces the same net result as

$$\begin{pmatrix} C & L & N & E & G \\ L & C & G & E & N \end{pmatrix},$$

for Charles' office passes at installation t to Evariste, then at installation s to Ludwig, Ludwig succeeds himself in office at installation t, then installs Charles at installation s, and so forth.   The theorem states that $s \circ (t \circ u) = (s \circ t) \circ u$; that is, if the composite mapping u-followed-by-t is followed by s, the net result is the same as u followed by the composite mapping t-followed-by-s.

The reason that all this can be compressed into the 11 words of the theorem is that several of those words have been ingeniously assigned just the right technical meanings, so that we can describe essential qualities exactly: mappings, associative (under), composition.

Next, we prove the theorem, following the line of argument suggested by the theorem itself when its language is expanded. Exactly what does the composite mapping $s \circ (t \circ u)$ do to a given element x in the set A? First the composite mapping $t \circ u$ maps x into some element of A, which we can label

$$(t \circ u)x.$$

But $t \circ u$ itself is a composite mapping made up of u followed by t. Suppose u maps the particular element x into an element $x^u \in A$. Then t maps x^u into an element $(x^u)^t = (t \circ u)x$. Finally s maps $(x^u)^t$ into a certain element $((x^u)^t)^s$ of A, and this is the element we call $s \circ (t \circ u)x$.

Suppose that at installation u the successions in the 5 offices are given by

$$u = \begin{pmatrix} C & L & N & E & G \\ L & E & C & G & N \end{pmatrix}.$$

What is the image of $C \in A$ under the composite mapping $s \circ (t \circ u)$? The composite mapping $t \circ u$ is made up of installation u, which maps C to L, followed by installation t, which maps L to L. Then $(t \circ u)C = (C^u)^t = L$. Finally, installation s maps L to C, showing that $((C^u)^t)^s = C$. We have followed the effect of the successive mappings on the one element C, but we could sum up the effects on all 5 elements:

$$s \circ (t \circ u) = \begin{pmatrix} C & L & N & E & G \\ N & C & G & L & E \end{pmatrix} \circ \left[\begin{pmatrix} C & L & N & E & G \\ E & L & N & G & C \end{pmatrix} \right.$$
$$\left. \circ \begin{pmatrix} C & L & N & E & G \\ L & E & C & G & N \end{pmatrix} \right] = \begin{pmatrix} C & L & N & E & G \\ C & E & L & N & G \end{pmatrix}.$$

Now we find what effect the other composite mapping, $(s \circ t) \circ u$ has on the same arbitrary element x of A that we mapped before. We adopted notation x^u for the map of x by u:

$$ux = x^u.$$

The effect of $(s \circ t)$ on x^u, since it is the composite of t followed by s, is the same as the effect of t operating on x^u and then s operating on the result. For the map of x^u by t we have the notation $(x^u)^t$:

$$tx^u = (x^u)^t.$$

After mapping by s we have

$$(s \circ t)x^u = s(x^u)^t = ((x^u)^t)^s,$$

the same element of A produced by $s \circ (t \circ u)$. Since x is perfectly arbitrary, we argue that for every element of A the mappings produce the same effect and so are equivalent as mappings of A. ∎

In the example, we have

$$(s \circ t) \circ u = \left[\begin{pmatrix} C & L & N & E & G \\ N & C & G & L & E \end{pmatrix} \circ \begin{pmatrix} C & L & N & E & G \\ E & L & N & G & C \end{pmatrix} \right]$$
$$\circ \begin{pmatrix} C & L & N & E & G \\ L & E & C & G & N \end{pmatrix} = \begin{pmatrix} C & L & N & E & G \\ C & E & L & N & G \end{pmatrix}.$$

The example we have been analyzing gives us a hint of the way groups will contribute to our study of Abel's Theorem. To prove that there is no formula that can give all the roots (solutions for x) of a general fifth degree equation, we shall assume that the roots *are* given by formula and study what happens if the roots are exchanged in various ways, much as the offices were exchanged at the installations. We will be able to show that all possible exchanges among the roots constitute a group under composition.

2. The integers I with $+$.   The integers are the "natural" or "counting" numbers $1, 2, 3, 4, \ldots$, their negatives $-1, -2, -3, \ldots$, and 0. Addition is a binary operation on the set of all integers, because for every pair of integers, i, j, there is exactly one sum integer, $i + j$.

To show that the system $I, +$ of the integers with addition as the binary operation form a group, we verify that $I, +$ satisfies the requirements of Definition 1–2:

i. For any three integers i, j, and k, $(i + j) + k = i + (j + k)$.

ii. The identity in I under $+$ is zero, because for any integer i, $i + 0 = 0 + i = i$.

iii. The inverse of a positive integer i under $+$ is $-i$; the inverse of 0 is 0; the inverse of a negative integer $-i$ is i.

In fact, the integers under $+$ form an abelian group, because for any integers i, j,

$$i + j = j + i.$$

Application.   Most of us are so count-conscious we find it hard to imagine primitive cultures with no counting words except "one," "two," and "many." We are even accustomed to frequent use of "zero," a much later invention. However, some may want to be reminded of applications for negative integers, such as debts *versus* credits, pounds lost *versus* pounds gained, number minutes early *versus* number minutes late, and so forth.

One of the most natural applications for all the integers is our system of credits and debts. A credit "balances" a debt of equal size, in the sense that they add to 0. The identity of this system is no transaction at all, the inverse of a gain is the corresponding loss, the inverse of a loss is the corresponding gain. Many a student who panics at abstract arithmetic with signed integers

$$+10$$
$$+ \quad -8$$
$$\overline{}$$

and mistakenly uses rules meant for multiplication ("A positive and a negative. Let's see. That makes a negative. -18.") can cope perfectly with the same problem if dollar signs are introduced:

$$+ \$10 \text{ gain}$$
$$+ \quad - \$8 \text{ loss}$$
$$\overline{}$$
$$+ \$2 \text{ net gain}$$

Even simple arithmetic calculations yield illustrations of associativity and commutativity. To add 27 and 36 we prefer something faster than counting 36 counts past 27. In effect we use something like the following, A showing which manipulations require associativity, C commutativity.

$$27 + 36 = (20 + 7) + (30 + 6) \overset{A}{=} 20 + [7 + (30 + 6)]$$

$$\overset{A}{=} 20 + [(7 + 30) + 6] \overset{C}{=} 20 + [(30 + 7) + 6]$$

$$\overset{A}{=} [20 + (30 + 7)] + 6 \overset{A}{=} [(20 + 30) + 7] + 6$$

$$\overset{A}{=} (20 + 30) + (7 + 6) = 50 + 13.$$

All this is involved implicitly when we add 27 and 36 in columns, adding the units digits to obtain 13, then the tens digits to obtain 50. To carry the 1 in 13, obtaining 63, requires further use of associativity.

To check addition of a column of figures, try associating the digits in a different way, so as not to repeat a mistake based on one number combination. Another bookkeepers' trick is to associate digits that add to multiples of 10 or of 5 or simply to bunch the digits conveniently when adding rather than adding them always one at a time.

3. The octic group—mappings of a square. There are eight (hence "octic") mappings of a square into itself that preserve distance between points, so that if points a and b are, say, 1.6 inches apart, then the distance between their images is also 1.6 inches. Adjacent vertices must map into adjacent vertices, center into center, and so forth, so that the only distance-preserving mappings are rotations of the square around its center, reflections (flips) about its various

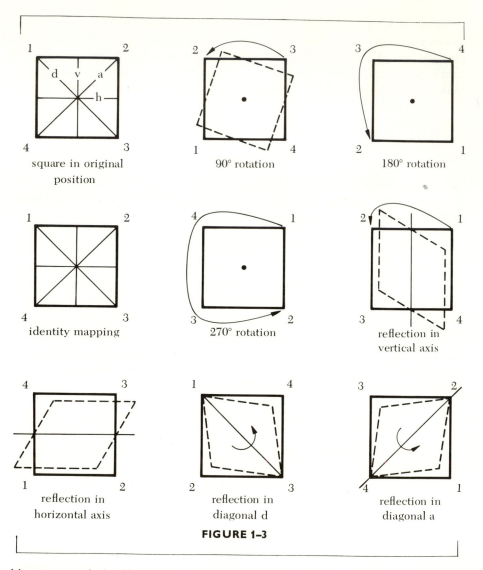

FIGURE 1–3

bisectors, and the identity mapping. You can study these mappings with a cardboard square labeled as in Figure 1–3.

Each mapping can be characterized by what it does to the vertices. For instance, a 90° counter-clockwise rotation could be characterized as

$$\begin{pmatrix} 1 & 2 & 3 & 4 \\ 4 & 1 & 2 & 3 \end{pmatrix}$$

because it moves vertex 1 into what was the 4 position, 2 into the 1 position, and so forth.

Exercise 1–18. Write similar characterizations for the other seven mappings.

Exercise 1–19.   Characterize the composite mapping $90° \circ d$; that is, show what happens to the vertices of the square when it is first reflected about d, then the resulting square rotated $90°$ counter-clockwise.

We can find the composite mapping $d \circ 90°$ by reference to the characterizations of the two mappings:

$$d \circ 90° \sim \begin{pmatrix} 1 & 2 & 3 & 4 \\ 1 & 4 & 3 & 2 \end{pmatrix} \circ \begin{pmatrix} 1 & 2 & 3 & 4 \\ 4 & 1 & 2 & 3 \end{pmatrix} = \begin{pmatrix} 1 & 2 & 3 & 4 \\ 2 & 1 & 4 & 3 \end{pmatrix}.$$

Let us follow the vertex 1, for instance. First, the $90°$ mapping maps 1 into 4, but then the d mapping maps that 4 into 2. Hence, the composite mapping maps 1 into 2.

Exercise 1–20.   Complete the composition table for the octic group:

	e	$90°$	$180°$	$270°$	d	a	v	h
e	e	$90°$		$270°$	d			
$90°$					$90° \circ d = h$			
$180°$		$270°$	e	$90°$			$180° \circ v = h$	
$270°$								
d	d	v						
a	a		d					
v		a						
h	h							e

Exercise 1–21.   Refer to Exercise 1–20 to prove that composition is a binary operation on the eight mappings of the square.

Exercise 1–22.   Spot check the composition table to verify associativity in at least five cases.

Exercise 1–23.   Prove that the eight mappings with the operation of composition form a group.

Exercise 1–24.   Show from the composition table that the group is not commutative.

Classroom Exercise 1–25.   This chapter contains six definitions. Read them aloud in unison.

The purpose of the preceding activity is to break down some of your reserve about pronouncing new words like "associativity." Mathematics teachers need all the techniques they can borrow from language teachers to help students with new vocabulary.

CHAPTER 2

OTHER ABSTRACT ALGEBRAS

In Chapter 1 we introduced an important abstract algebra, the group, and showed that we have in the integers with binary operation + a familiar example. The group postulates can, indeed, be thought of as abstracted from various properties of the integers. However, the group properties under addition do not exhaust the possibilities offered by the integers, which have a structure with respect to a second operation, multiplication, and even a distributive property linking + and ·.

In this chapter we introduce several abstract algebras—rings, integral domains, and fields—and show that their various postulates can be abstracted from the properties of various number systems—the integers, the rational numbers, the real numbers, and the complex numbers.

Finally, with enough examples to make the concept meaningful, we define "an abstract algebra."

Definition 2–1. A **ring** R is a set of elements with two binary operations, addition (+) and multiplication (·), satisfying three postulates:

i. R is an abelian group under the operation +.
ii. Multiplication is associative in R.
iii. The **right-** and the **left-distributive laws** hold; that is, for elements

r, s, and t of R,

$$r \cdot (s + t) = r \cdot s + r \cdot t \quad \text{and} \quad (r + s) \cdot t = r \cdot t + s \cdot t.$$

We assume R has elements in it, but only one is enough, for even the system composed of the one element 0 with addition and multiplication tables

satisfies the requirements and so is a ring. The operations are really arbitrary, although we use the ordinary symbols $+$ and $\cdot$ to suggest the prototype of rings, the ring of integers. The symbols $+$ and $\cdot$ can be used to denote any two binary operations that fit the requirements (such as the distributive laws).

Since the ring of integers also satisfies some other requirements, we introduce some abstract algebras satisfying more restrictions than groups or rings.

Definition 2–2. A **ring with unit element** is a ring that has a multiplicative identity e, *i.e.*, an element e such that $r \cdot e = e \cdot r = r$ for every ring element r.

Definition 2–3. A **commutative ring** is a ring in which multiplication is commutative.

Definition 2–4. An **integral domain** is a commutative ring with unit element with **no proper divisors of zero**; that is, a product $a \cdot b$ cannot equal zero unless one of the factors a or b is zero.

Definition 2–5. A **field** is a ring with at least two elements whose nonzero elements form an abelian group under ·. Alternatively, we could say, "A field is a commutative ring with unit element containing a multiplicative inverse for each nonzero member."

These definitions are more restrictive than Definition 1–2 of groups; that is, they impose more than the group requirements. One abstract algebra with fewer restrictions than a group is the semigroup. It is important to mathematicians and receives study even though it has no identity nor inverse properties.

Definition 2–6. A **semigroup** is a set with an associative binary operation.

Examples of Various Abstract Algebras

1. The natural numbers N. How do we know the facts of arithmetic? Have we any knowledge beyond limited experience of properties like associativity and distributivity? Are there deductive proofs that show these laws always hold, even for very large numbers we have not tried? An honest answer would be a qualified "yes."

Definition 2–7. The **natural numbers N**, or **positive integers**, are the counting numbers

$$1, 2, 3, 4, 5, \ldots,$$

not just the "numerals," which are only a set of marks, but the numbers themselves with all their familiar properties of order, succession, and so forth.

We can base multiplication in N on Cartesian products of disjoint sets, as in the following illustration. To find a product like $2 \cdot 3$ we start with a set of two objects, say two hands, and assign to each a set of three objects, say three fingers. We can demonstrate commutativity by starting with three objects and

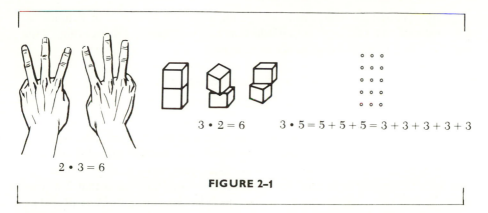

$$3 \cdot 2 = 6 \qquad 3 \cdot 5 = 5 + 5 + 5 = 3 + 3 + 3 + 3 + 3$$

$$2 \cdot 3 = 6$$

FIGURE 2–1

assigning to each a set of two (Fig. 2–1). We can base multiplication on addition by saying that to multiply a times b we add a summands, each equal to b (or add b a's).

We can base addition on counting sets. We count $2 + 3$ by counting members in the union of a set with 3 members and a separate set with 2 members.

When we go back one more step and try to define counting, or the counting numbers N, we have to be careful to avoid circularity. For instance, if we defined the "next" or "immediate successor" number n^+ of a number n to be $n + 1$, we would have a circular definition, because addition is based on counting. We simply add "next" or "successor" to our list of commonly understood but mathematically undefined terms. Such important basic ideas as "set," "point," "line," and "equally likely" are on this list of undefined terms. So is the idea of 1 as a first, or smallest, counting number.

Definitions in mathematics, as we have seen in the case of groups, are rather arbitrary. It is possible to define "1" or "successor" in terms of other counting concepts, but some nucleus of basic ideas must remain undefined.

One property of the counting numbers N usually taken as part of their definition is mathematical induction.

Definition 2–8. The **mathematical induction property** of the counting numbers N can be stated: If a subset S of N contains 1 and contains the successor of each of its members, then it contains (and equals) all of N.

$$2 + 3 = 5 \qquad 3 + 2 = 5$$

FIGURE 2–2

This property is used again and again in proofs and also in definitions. For instance, we said loosely that addition can be "based" on counting. This can be done precisely by an *inductive definition*

$$\text{addition } (+) \text{ in } N \begin{cases} n + 1 = n^+, \text{ the successor of } n \in N, \\ n + m^+ = (n + m)^+. \end{cases}$$

This shows how to find sums once you know how to count, and it does so by an inductive rule. That is, first it shows how to add 1 to things; then it shows how to add the next larger number m^+ to things, if you know how to add m. To add 2 and 2 by this definition, for instance, we would first learn that $2 + 1 = 2^+ = 3$, and then, using 1 as m, add $m^+ = 2$ by the rule $2 + 1^+ = (2 + 1)^+$, which, since $2 + 1 = 3$, is 3^+, or 4.

Classroom Exercise 2-1.   Find $4 + 3$ directly from the preceding definition, by finding $4 + 1$, $4 + 2$, and so forth, assuming you know only how to count. Act it out with objects, such as blocks or pebbles.

Multiplication in N can be defined inductively in terms of addition, as

$$\text{multiplication } (\cdot) \text{ in } N \begin{cases} (i) \ 1 \cdot n = n \\ (ii) \ (m + 1)n = mn + n \end{cases}$$

(we drop the dot for convenience).

As a sample of proof by mathematical induction, we prove that in N multiplication is distributive on the left with respect to addition. We will assume that we have already satisfied ourselves (by proofs) that addition is associative and commutative in N. Note the names of the parts of this formal proof pattern. They suggest the way to construct such a proof yourself.

Theorem 2-1.   If k, m, and n are in N, then

(*iii*) $k(m + n) = km + kn$.
Proof by induction on k:

Basis.   Does (*iii*) hold when k takes on the special value 1? We form $k(m + n)$ at the value $k = 1$:

$$k(m + n)\big|_{k=1} = 1(m + n)$$

$$= m + n, \text{ by } (i) \text{ in the definition of multiplication,}$$

and

$$km + kn\big|_{k=1} = 1m + 1n$$

$$= m + n, \text{ by two applications of } (i).$$

Comparing, we find that (*iii*) holds for $k = 1$.

Now we build up a subset S of N made up of all those values of k that make (iii) hold. We have just proved in the basis that S has at least one member, the number 1.

Induction Step

Induction Hypothesis: Assume some number s is a member of S; that is, (iii) holds when k takes on the value s, or

$$k(m + n)\big|_{k=s} = s(m + n) = sm + sn = km + kn\big|_{k=s}.$$

Then is the successor $s^+ = s + 1$ of s automatically a member of S? We have to prove that

$$k(m + n)\big|_{k=s+1} = km + kn\big|_{k=s+1}.$$

We have

$$k(m + n)\big|_{k=s+1} = (s + 1)(m + n) = s(m + n) + (m + n), \text{ by}$$
$$(ii),$$
$$= (sm + sn) + (m + n), \text{ by the induction}$$
$$\text{hypothesis,}$$
$$= (sm + m) + (sn + n), \text{ by commutativity and}$$
$$\text{associativity of } +,$$
$$= (s + 1)m + (s + 1)n, \text{ by } (ii),$$

completing the proof. ∎

Definition 2–9. A **proof by mathematical induction** has two parts, the **basis,** in which we verify that 1 is in the subset S of N for which some property holds, and the **induction step.** In the induction step we assume the **induction hypothesis** that s is a member of S and prove that $s + 1$ is in S.

How does this prove the theorem? We cannot prove that mathematical induction proves theorems. We assume that it does in assuming the induction property of the natural numbers. We can enlarge enough on the idea to make it seem plausible, however. In the basis of the induction in Theorem 2–1 we showed that formula (iii) works for $k = 1$. In the induction step we showed that if it works for $k = s$ then it works for $k = s + 1$. Among others, we can take 1 for our s, and say, "It works for 1, as verified in the basis, so it works for 1^+, or 2." Then we can take 2 as s and say, "It has been proved to work for 2 so it works for 3." The induction step proves that we can go from any previously established step to the next one.

This process has been compared to an infinite "topple the dominos" game. Suppose you have set up a long line of dominos close enough together and well enough aligned so that each domino is guaranteed in falling to knock over the next one (induction step). Suppose you push over the first one (basis). Then it seems that all the dominos will fall. Sometimes it seems likely that the whole process will be illustrated some day by multiple collisions on a freeway.

Another popular illustration is that of an infinitely tall ladder. You establish a basis when you step on the first rung. You use the induction step for climbing from whatever rung you have reached to the next higher one.

Classroom Exercise 2-2.   What part of the mathematical induction fails in the ladder illustration if the tenth rung is broken or missing? if the first rung is too high above the ground for you to reach? if you are afraid of heights above 15 feet?

Classroom Exercise 2-3.   Set up a display of real dominos and use them to explain the idea of mathematical induction.

Exercise 2-4.   Prove that N is a semigroup with binary operation $+$. Use mathematical induction on c to prove the associative law, $a + (b + c) = (a + b) + c$. You will need several applications of the inductive definition of addition. Which group properties fail to hold in the semigroup N?

Classroom Exercise 2-5.   Have a classmate hold up his fingers separated into three bunches. Show how to demonstrate for a class how associativity of addition follows from counting. Demonstrate commutativity of addition.

Exercise 2-6.   Prove that N is a semigroup with multiplication as binary operation. Assuming the distributive law, prove by induction on k that $(km)n = k(mn)$. What group properties fail to hold in this case?

Exercise 2-7.   Prove that if A is a member of a semigroup with operation θ, then A commutes with powers of A; that is, $A \, \theta \, A^n = A^n \, \theta \, A$. This result must be proved inductively on n, for until we know, for instance, that $A \, \theta \, A^2 = A^2 \, \theta \, A$, the power A^3 is not well defined (that is, A^3 would be ambiguous if $A \, \theta \, A^2 \neq A^2 \, \theta \, A$).

The properties of N, $+$, $\cdot$ can be postulated at the outset as assumed features of arithmetic. Alternatively, a set of postulates such as those of Peano can be adopted along with inductive definitions of $+$ and $\cdot$, and the properties can then be deduced by mathematical induction. (See C. C. MacDuffee, *Introduction to Abstract Algebra*, Wiley, 1940, pp. 1–5; or E. Landau, *Foundations of Analysis*, Chelsea, 1960.)

Countable Infinity

The natural numbers taken as a whole have the peculiar property that they can be put into one-to-one correspondence with a proper subset of themselves. (By a "proper" subset we mean a set that has some but not all members of the original set.) A classic illustration of the peculiarities of arithmetic in such a case is that of the hotel with rooms numbered 1, 2, 3, . . . , one room for each natural number. Suppose every room is full, but another guest arrives. The manager simply gives him room No. 1, whose former occupant moves to No. 2, whose former occupant moves to No. 3, and so on. What would be an impossible problem in a finite hotel is solved easily in a *countably infinite* hotel.

> **Definition 2–10.** A set is called **countably infinite** if its members can be put into one-to-one correspondence with the natural numbers. The set is then said to have **cardinal number** $\aleph_0$ (read "a'-leph null," from the first letter of the Hebrew alphabet). A finite set of n members has **cardinal number n** and is also said to be countable.

2. The integers I. The natural numbers N form a semigroup under $+$ (see Exercise 2–4) but not a group. The integers I, 0, 1, -1, 2, -2, 3, . . . , represent the completion of N under subtraction; that is, they include the additive identity and inverses, so that subtraction, the inverse of addition, becomes a binary operation on I.

If 0 is to serve as the additive identity, it has to be the solution for $\aleph_0$ problems:

$$1 + x = 1, \quad 2 + x = 2, \quad 3 + x = 3, \ldots,$$

(see Definition 1–2 *ii*). One way of adjoining an additive identity to N is to lump all these problems together in an "equivalence class" and equate the answer to the problems: Dropping all but the pair of numbers (a, b) in the problem $a = b + x$, we write the equivalence class as

$$\{(1, 1), (2, 2), (3, 3), \ldots\}$$

and call it 0 for short.

We have answers in N for many problems $a = b + x$, such as $10 = 5 + x$, or $(10, 5)$, for $(12, 2)$, for $(3, 2)$, and so on. We lump the problems into classes, calling pairs equivalent if they stand for problems having the same answer. Then we refer to them by the answer itself, as

$$5 \sim \{(6, 1), (7, 2), (8, 3), (9, 4), \ldots\}.$$

We could define operations $+$ and $\cdot$ for these equivalence classes so that the resulting sums and products would correspond to the sums and products in N. The equivalence classes and the operations of their arithmetic would look quite different from N, but later we shall introduce a word "isomorphic" to describe the fact that they are arithmetically alike.

Now the fact that we do not have all the answers in N for problems $a = b + x$ with a and b in N (for instance, for $3 = 7 + x$) does not seem so bad, for at least we have the problems! After improving their description somewhat, we could do arithmetic with whole classes like

$$\{(1, 4), (2, 5), (3, 6), \ldots\},$$

simply "playing 'em as if we had 'em," as one professor used to say. This is a technique characteristic of algebra, after all, where we pretend we know the answer and temporarily call it x while we arrange facts known about it. It is more convenient to pin labels on the equivalence classes, so we will not need to write them out. We choose $-n$ as an appropriate label for the class having $(1, 1 + n)$ for one of its members.

> There is precedent in medical diagnosis for this idea of equating the answer to the problems. The doctor lumps together all individuals with a certain list of symptoms—sore throat, spots before the eyes, palsied left great toe—and labels the file "Max's Disease." He may not have cured the disease, but he has defined it.

We have an immediate need for precise vocabulary at this point to clarify the general notion of equivalence classes.

Equivalence and Other Relations

We have been lumping together all the $\aleph_0$ number pairs (x, y) that satisfy $x = y + h$ and calling them "equivalent." We have built up to this form of equality so as to make it appear reasonable; yet our adoption of it is quite arbitrary. Does it need any justification? Does an "equivalence" relation have to obey any rules? For that matter, precisely what is a "relation?"

> **Definition 2–11.** A **relation** on a set S to a set T is a subset of the Cartesian product set $S \times T = \{(s, t)$, where $s \in S$, $t \in T\}$.

This technical definition is easy to understand if you already know what a relation is, but it is otherwise not very illuminating. That is why it does not occupy the first lines of this book, a more logical position for it. Consider the familiar relation "less than" or "$<$" applied to real numbers. Here both S and T are taken to be R, the real numbers. We say that a member (r_1, r_2) of the Cartesian product set $R \times R$ is in the relation set "$<$," if $r_1 < r_2$. Then $(1.3, 5)$, $(-\sqrt{10}, -0.2)$, and $(0, 0.0001)$ are in the relation $<$, while $(3, 2)$, $(-0.2, -0.21)$, and $(9, 9)$ are not.

> If $\bar{R}$ is a relation, then it is convenient to write $\mathbf{s} \, \bar{\mathbf{R}} \, \mathbf{t}$, "$s$ is related to t by the relation $\bar{R}$" if $(s, t) \in \bar{R}$.

> **Definition 2–12.** A **mapping** from a set S to a set T is a relation on S to T such that each member of S is related to exactly one member of T. A mapping is **onto** if each member of T is related to at least one member of S.

Since a binary operation is a mapping (Definition 1–1), it can be defined in terms of relations, too, as a relation on the set $S \times S$ to the set S. For instance, the operation addition in N, thought of as a relation, is the set of elements $[(m, n), m + n]$, where m and n are in N. It contains $((2, 3), 5)$ and $((17, 1), 18)$ but does not contain $((1, 1), 1)$ or $((m, n), 1)$ for any m and n in N.

Let us mention again that these definitions of relation, mapping, and operation are very formal. They do not really explain the concepts, but they do make our vague ideas into precise instruments we can use exactly.

A few examples may point up the versatility of the mapping idea: (1) A newscaster chalks the temperature beside each major city on a weather map. This gives us a mapping from cities to temperatures. On a given day not all possible temperatures would be likely to appear, so the mapping would be into, not onto, the possible temperatures. Ordinarily there would not be an inverse mapping from temperatures to cities, as several cities would probably have the same temperature. (2) A listing of batting averages for baseball players gives a similar "into" mapping, ordinarily without an inverse mapping. (3) The polynomial function $y = x^2$ gives a mapping from the real numbers x into the real numbers. The mapping is onto the nonnegative real numbers, but there is no inverse mapping from the nonnegative real numbers y to the real numbers x, since both $+x$ and $-x$ have the same square. (4) A chart gives the wind speed and direction at 100 locations in a wind tunnel. Each location is given by three coordinates. The chart provides a mapping from the locations to the speeds and from the locations to the directions.

> **Definition 2–13.** An **equivalence relation** on a set S is a relation E on S to S that is
>
> *i.* **reflexive:** $s \, E \, s$ for every $s \in S$.
> *ii.* **symmetric:** If $s \, E \, t$, then $t \, E \, s$.
> *iii.* **transitive:** If $s \, E \, t$ and $t \, E \, u$, then $s \, E \, u$.
>
> The elements s, t, and u are elements of S; note that every element of S is involved in requirement i.
> All elements of S equivalent to an element s form an **equivalence class.**

You have worked with many different equivalence relations, such as congruence and similarity in geometry, equality of ratios $\frac{4}{6} = \frac{2}{3}$, equality of angles differing by multiples of $360°$, and so on.

Classroom Exercise 2–8. Discuss: A class of 267 students S are given assigned desks D in a 300-desk classroom. On an exam each makes some numerical grade G from 0 to 100. Does the correspondence between students S and desks D constitute a mapping from S into D? From S onto D? from D into S? Does the correspondence between students and grades constitute a mapping from S into G? S onto G? from G into S? Is "got-the-same-grade-as" an equivalence relation on S?

> We often use the fact that an equivalence relation $\sim$ on a set S partitions S into mutually exclusive and exhaustive **equivalence classes:** each equivalence class is made up of an element s of S together with all elements equivalent to it. The classes are mutually exclusive, for if classes C and D have an element s is common, then all elements of S that are equivalent to s lie in C and in D, so that $C = D$. Together the equivalence classes exhaust S, because the reflexive property allows us to consider a single element s an equivalence class.

Notice that, conversely, any partition of S into mutually exclusive and exhaustive subsets leads to an equivalence relation on S, that is, the relation $s \sim t$ if and only if s and t are members of the same subset.

Exercise 2–9. Take it as known that N under $+$ is associative and commutative and has the cancellation property $m + d = n + d$ implies $m = n$. Show that if we say $(x, y) \sim (z, w)$ whenever $x + w = y + z$, $x, y, z, w \in N$, then $\sim$ is an equivalence relation on the pairs of numbers (x, y) from N. Notice that this is the equivalence used in our subtraction-problem classes.

As we have anticipated in Exercise 2–9 we can prove that all the pairs (x, y) of numbers from N that satisfy $x = y + h$ are equivalent, where we define pairs (x, y) and (z, w) to be equivalent whenever $x + w = y + z$. Similarly, the pairs $(1, 1 + k)$, $(2, 2 + k)$, $(3, 3 + k)$, ... that satisfy $x + k = y$ are equivalent. The pairs $(1, 1)$, $(2, 2)$, $(3, 3)$, ... that satisfy $x = y$ are equivalent. Then we finally have justification for our procedure of lumping number pairs together as "equivalent".

We now drop the equivalence-class notation, replacing the class containing $(1 + n, 1)$ by n, the class containing $(1, 1)$ by 0, and the class containing $(1, 1 + n)$ by $-n$. An arithmetic can be defined for the integers that is compatible with the arithmetic of N for positive integers. It can be proved that with this arithmetic of signed integers the integers I form a ring.

Definition 2–14. The **ring of integers I** is made up of the positive integers $N = 1$, 2, 3, ..., their negatives (or additive inverses) -1, -2, -3, ..., and 0, with addition and multiplication as the ring operations. The additive identity is 0; the multiplicative identity is 1.

Classroom Exercise 2–10. Prove that, in a ring R, $m \cdot 0 = 0$ for each $m \in R$. (Write $m \cdot 0 = m(0 + 0)$, apply the distributive law, and add like quantities to both sides.)

Classroom Exercise 2–11. As isolated unrelated rules, the arithmetic of signed numbers can be puzzling and discouraging to students. Show how to explain, as to a baffled algebra student, why a positive integer times a negative yields a negative but a negative times a negative yields a positive. (Write $m(n - n)$, use the distributive law and Classroom Exercise 2–10.)

Since multiplication is commutative, I satisfies the requirements of Definitions 2–2 and 2–3 and so is a commutative ring with unit element. Does I satisfy Definition 2–4 also?

Theorem 2–2.   The integers I form an integral domain.

PROOF: Suppose the product ab of two integers a and b is 0, but $a \neq 0$ and $b \neq 0$.   Since $ab = 0$, $-(ab)$ also equals 0, so that each of the four products ab, $a(-b)$, $(-a)b$, and $(-a)(-b)$ equals zero.   One of them is the product of two positive integers in N, since $a \neq 0$ and $b \neq 0$; for instance, if $a \in N$ but $-b \in N$, choose the product $a(-b)$.   However, multiplication is a binary operation in N, so N is closed under multiplication.   This means that the product of two numbers of N lies in N.   Then the product cannot be 0.   ∎

> This argument illustrates an important proof technique, called the *reductio ad absurdum* proof or proof by *contradiction*.   We show that if the theorem failed to hold we could draw conclusions we have already shown are not valid; that is, we "reduce the case to an absurdity."

How many integers are there?   Since I includes all $\aleph_0$ members of N, their $\aleph_0$ respective negatives, and 0, we might say that I contains $2\aleph_0 + 1$ integers.   However, we have a convention in counting that two sets have the same "cardinal number" or "count" if there is a one-to-one correspondence between their members.

Exercise 2–12.   Consider the integers in the arrangement 0, 1, -1, 2, -2, 3, -3, ...   Let $n \in N$.   Which integer occupies the $2n$ position? the $2n + 1$ position?   Prove that I and N can be put into one-to-one correspondence.

From Exercise 2–12, I has $\aleph_0$ integers.   As we have seen, adding one new member to a countable infinity does not change the cardinal number of a set.

Exercise 2–13.   Prove that adding a finite number n of new members to a countably infinite set does not change its cardinal number.   (Find a one-to-one correspondence as in the hotel illustration preceding Definition 2–10.)

Exercise 2–14.   Prove that n countably infinite sets can be put into one-to-one correspondence with one countably infinite set.   (Use the technique of Exercise 2–12.)

3. The rational numbers Q.   By introducing the integers we made it possible to solve equations of the form

$$x + b = a, \quad \text{where } a, b \in N.$$

Now we look at equations of the form

$$bx = a, \quad \text{where } a, b \in I.$$

For some number pairs (a, b) there are solutions x in I; for instance, if $(a, b) = (6, 2)$, $x = 3$ is a solution. If $(a, b) = (18, -3)$, then $x = -6$ is a solution. If $(a, b) = (51, 17)$, $x = 3$ is a solution.

Following the same number-pair approach used to introduce the integers, we form equivalence classes of number pairs (a, b) that have a common solution x for $bx = a$. For example, the following are equivalence classes:

$$\{(2, 1), (4, 2), (-8, -4), \ldots\}$$
$$\{(9, -3), (-6, 2), (-15, 5), \ldots\}.$$

How are (a, b) and (c, d) related if they have the same solution s?

$$bs = a \text{ implies } d(bs) = d(a), \text{ so that } bds = ad,$$

and

$$ds = c \text{ implies } b(ds) = b(c), \text{ so that } bds = bc,$$

so that we have

$$ad = bc.$$

Exercise 2–15.   Write four number pairs equivalent to $(4, 1)$.

Exercise 2–16.   Prove that if we write $(a, b) \sim (c, d)$ whenever $ad = bc$, then $\sim$ is an equivalence relation on ordered pairs (x, y) with $x, y \in I$ and $y \neq 0$. Take it as known that I under multiplication is associative and commutative and has the cancellation property $ac = bc$ implies $a = b$ for any $c \neq 0$.

When number pairs (a, b) and (c, d) have solutions in I in the sense that there are integers $r, s \in I$ for which $br = a$ and $ds = c$, we call the pairs equivalent if $ad = bc$. Since we have shown in Exercise 2–16 that this equivalence extends to pairs that have no solutions in I, so long as $b \neq 0$ and $d \neq 0$, we are justified in working with equivalence classes of such pairs.

We leave out the case $b = 0$, because our purpose is to solve $bx = a$. If $b = 0$, $bx = 0$ even for large x's, so we cannot solve $0x = a$ unless $a = 0$, also. What happens if a does equal 0? Then $0x = 0$, no matter what x we choose, so the solution is underdetermined.

Again we equate the answer to the class of problems, as

$$\tfrac{3}{1} \sim \{(3, 1), (6, 2), (-12, -4), \ldots\}$$
$$\tfrac{2}{3} \sim \{(2, 3), (-4, -6), (10, 15), \ldots\}$$

$$\frac{a}{b} \sim \{(a, b), a, b \in I, b \neq 0 \text{ and all pairs } (c, d), d \neq 0, \text{ for which } ad = bc\}$$

We call these answers "rational numbers," "rational" here being simply an adjectival form of "ratio."

We can define addition and multiplication of these rational numbers so that the results will be compatible with the arithmetic of I. It would be simpler just to add numerators and add denominators when adding fractions, without the bother of common denominators, but it would lead to situations like

$$\tfrac{6}{2} + \tfrac{6}{3} = \tfrac{12}{5},$$

which would not agree with our experience with the integers 3 and 2.

If r is a solution in I to $bx = a$ and s is a solution in I to $dx = c$, then $r + s$ is a solution to $(bd)x = ad + bc$, and rs is a solution to $(bd)x = ac$.

Exercise 2–17.   Establish the result in the last sentence.

We can now define

$$\frac{a}{b} + \frac{c}{d} = \frac{ad + bc}{bd}$$

$$\frac{a}{b} \cdot \frac{c}{d} = \frac{ac}{bd}$$

and we extend these definitions to apply to any number pairs (a, b), $a, b \in I$, $b \neq 0$, even if there is no solution in I for $bx = a$. Because these definitions are based on solutions of equations $bx = a$, just as our definition of equivalence was, the sums and products are independent of which elements of the equivalence classes are used. For instance, if we add or multiply $\tfrac{2}{3}$ and $-1/2$, we can be sure of obtaining the same results as if we used their equivalents $-4/-6$ and $-5/10$.

Soon we shall introduce "isomorphism" (Definition 2–15) and show that the equivalence classes containing pairs $(a, 1)$ with the arithmetic of rational numbers are essentially (here, arithmetically) like I.

Exercise 2–18.   Form the sum and the product of the fractions 7/8 and $-5/12$. Select equivalents for the two rational numbers and form their sum and product.

Exercise 2–19.   Show that a/b and $an/(bn)$, where $a, b, n \in I$, $b \neq 0, n \neq 0$, are in the same equivalence class of rational numbers.

Exercise 2–20.   Show that $a/b + c/d$ and $an/(bn) + c/d$, $n \in I$, are equivalent. All denominators are assumed to be nonzero.

Exercise 2–21.   Show that $(a/b)(c/d)$ and $(an/(bn))(c/d)$, $n \in I$, are equivalent. All denominators are assumed to be nonzero.

Theorem 2–3.   The rational numbers $Q: \{a/b, a, b \in I, b \neq 0\}$ form a field with operations $+$ and $\cdot$.

Proof: The sum of a/b and c/d is $(ad + bc)/(bd)$. From Theorem 2–2, $bd \neq 0$. Because multiplication is an operation on I, bd, ad, and bc are all uniquely defined members of I. Because addition is an operation on I, $ad + bc$ is some definite member of I. Then the sum has numerator and denominator in I and the denominator is not 0, and therefore the sum is in Q, uniquely defined as the equivalence class for the numerator-denominator pair.

Associativity of $+$ in Q is "inherited" from I because of the associativity of $+$ and $\cdot$ in I and the distributive property of $\cdot$ with respect to $+$, as we verify by direct computation:

$$\left(\frac{a}{b} + \frac{c}{d}\right) + \frac{e}{f} = \frac{ad + bc}{bd} + \frac{e}{f} = \frac{(ad + bc)f + (bd)e}{(bd)f}$$

and

$$\frac{a}{b} + \left(\frac{c}{d} + \frac{e}{f}\right) = \frac{a}{b} + \frac{cf + de}{df} = \frac{a(df) + b(cf + de)}{b(df)}.$$

Exercise 2–22. Prove that Q under $+$ has an identity, inverses, and the commutative property.

Exercise 2–23. Prove that $\cdot$ is a binary operation on Q.

Exercise 2–24. Prove that $\cdot$ is associative on Q.

Exercise 2–25. Prove that $Q - \{0\}$, that is, Q with 0 excepted, under $\cdot$ has an identity, inverses, and the commutative property. Prove that in Q $\cdot$ is distributive with respect to $+$.

The exercises complete the proof of Theorem 2–3. ∎

Classroom Exercise 2–26. A clerk computes a 5 per cent sales tax on each customer's bill. She charges Tom 1 cent tax on a 10-cent pencil, Dick 1 cent tax on a 10-cent pencil, but Harry only 1 cent tax on two 10-cent pencils. Which field requirement is violated?

Isomorphism

The subset Q_1 of Q made up of all numbers of Q that can be written with denominator 1 is essentially like I. By "essentially like I" we mean "like I in the respects that are important to us here." We often have a use in everyday life for the notion of "essentially alike but not identical." We use it whenever we point out similarities between nonidentical things, as when we say things are "alike for all practical purposes." Mathematical language has been developed for making these comparisons precise and stating their limits specifically.

You will not need to apply this first definition in all its generality, but read it to get the gist of it.

Definition 2–15.   Let S be a system of elements interrelated by a relation Ⓡ . Ⓡ can be a relation on S to S, a binary operation on S, a relation on $S \times S \times S$ to S, and so forth. Let T be another system of elements interrelated by a relation Ⓡ of similar type (unary, binary, ternary, or other). S and T are **isomorphic** with respect to their respective relations Ⓡ and Ⓡ, written $\mathbf{S} \cong \mathbf{T}$, if there is a one-to-one correspondence between their elements that induces a one-to-one correspondence between the subsets that belong to their respective relations.

We shall be using the simpler definition of isomorphism that applies when the relations are binary operations.

Let a binary operation $*$ be defined on S and a binary operation $\circ$ be defined on T. Then $\mathbf{S} \cong \mathbf{T}$ **with respect to the operations** means that there is a one-to-one correspondence between all members of S and all members of T that induces correspondence between the results whenever the operations are applied to corresponding elements (Figure 2–3).

Mathematical isomorphism is an especially useful kind of "equality," since it can be delimited exactly. To the census taker, family S may "equal" family T because each consists of father and mother and three children, but S and T may be far from equal medically, socially, politically, and so forth. For some purposes it may be useful to regard schools S and T as "equal" if each has grades one through six. For other purposes they might be "equal" if they had the same teacher-pupil ratio and so forth. The term "isomorphism" lets us say exactly in what respects two systems are alike.

Theorem 2–4.   The field Q contains a subring Q_1 isomorphic to I with respect to $+$ and $\cdot$.

PROOF: Let Q_1 be made up of the numbers of Q that can be written with denominator 1: $\frac{3}{1}$, $-17/1$, $\frac{0}{1}$, $\frac{1}{1}$, and so forth. Set up the one-to-one correspondence

$$\frac{a}{1} \leftrightarrow a \tag{1}$$

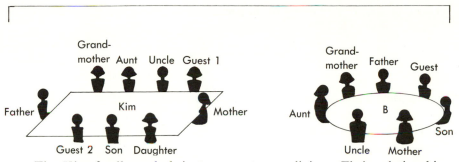

The Kim family and their two guests are dining. Their relationships, geneological and otherwise, are shown by the labels. The Burbach family and a guest are playing cards.

Let the subset K of Kims correspond to the Burbachs, as follows:

K	B		K	B
Father $\leftrightarrow$	Father		Mother $\leftrightarrow$	Mother
Grandmother $\leftrightarrow$	Grandmother		Son $\leftrightarrow$	Son
Aunt $\leftrightarrow$	Aunt		Guest 2 $\leftrightarrow$	Guest
Uncle $\leftrightarrow$	Uncle			

This correspondence preserves the relationships among the subset K of Kims and the Burbachs, respectively.

Now the Kims start passing dishes, yielding a mapping m, where k passes a dish to k^m. Father passes the cucumbers to Grandmother ($F^m = GM$), Uncle passes the chicken around Guest 1 to Mother ($U^m = M$), $GM^m = A$, $A^m = U$, $M^m = S$, $S^m = G_2$, $G_2^m = F$, so that m maps the subset K of Kims onto itself. In their card game each of the Burbachs passes a playing card to the player on his right, yielding a mapping d ($F^d = GM$, $U^d = M$, and so forth). The correspondence preserves the mappings m and d, respectively.

The mappings that interest us especially in arithmetic are binary operations, which we represent schematically:

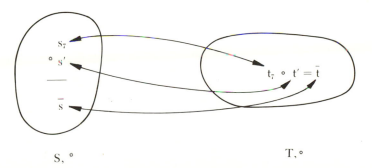

If the problems correspond, then the answers correspond; If s_7 and s' correspond respectively to t_7 and t', then $s_7 * s' = \bar{s}$ corresponds to $\bar{t}$, or $t_7 \circ t'$.

FIGURE 2–3

between Q_1 and I. According to the operation $+$ defined on Q, we have

$$\frac{a}{1} + \frac{b}{1} = \frac{a \cdot 1 + 1 \cdot b}{1 \cdot 1} = \frac{a + b}{1}.$$

Then, adding a and b within I according to the definition of $+$ in I, we obtain $a + b$. The two sums then do correspond according to (1).

The product as defined in Q is

$$\frac{a}{1} \cdot \frac{b}{1} = \frac{a \cdot b}{1 \cdot 1} = \frac{ab}{1},$$

which corresponds according to (1) to the product ab as defined in I. This shows that the correspondence is an isomorphism and so justifies also the assertion that Q_1 is a ring, since all the ring properties are properties of $+$ and $\cdot$, which Q_1 has been shown to share with I, a known ring. ∎

Exercise 2–27. Let Q_2 be the rational numbers with denominator 2. Prove that $Q_2 \cong I$ with respect to $+$ as defined in Q and in I, respectively.

We can show, as in Appendix A, that Q has $\aleph_0$ members.

Classroom Exercise 2–28. Let C be the "clock group" with elements 3, 6, 9, and 0 hours, operation ($+$) addition with the convention that 12 hours $= 0$ hours on the clock. Let G be the group with operation $*$ described in the operation table:

$*$	1	r	s	t
1	1	r	s	t
r	r	s	t	1
s	s	t	1	r
t	t	1	r	s

Let H be the Klein four-group with operation θ.

θ	1	a	b	c
1	1	a	b	c
a	a	1	c	b
b	b	c	1	a
c	c	b	a	1

Prove that two of the three groups are isomorphic, but that the third is not isomorphic to them.

4. The real numbers R. We can find solutions x for $bx = a$, $b \neq 0$, in the rational field Q, if a and b are integral or rational. However, it has been recognized for centuries that Q cannot supply solutions to many equations of higher degree. For instance, the following result and its proof were known to Greek mathematicians over 25 centuries ago. It is based on prime divisors, but only for the prime 2, so we will assume enough knowledge about evenness and oddness of integers to make the proof accessible, although our study of prime integers comes in Chapter 5.

Theorem 2–5. There is no rational solution x for the equation

$$x^2 = 2.$$

PROOF: Suppose $a/b \in Q$ is a solution for $x^2 = 2$; that is,

$$\left(\frac{a}{b}\right)^2 = 2.$$

Suppose a and b have no common factor, so that a/b is in lowest terms. In particular, the proof hinges on the assumption that a and b are not both even. Multiplying both sides of the equation by b^2, we have

$$a^2 = 2b^2.$$

Then a^2 and $2b^2$ must stand for the same integer n. The integer n is not zero, since $a/b \in Q$, so that $b \neq 0$. From $n = 2b^2$, we see that n is an even integer, a multiple of 2. Then $n = a^2$ must be even also. Then a must be even, since an odd number squared would be odd ($(2m + 1)^2 = 4m^2 + 4m + 1$). Since a is even, represent it by $2k$, where $k \in I$. Then

$$n = a^2 = (2k)^2 = 4k^2.$$

But if n is a multiple of 4, $2b^2$ must be a multiple of 4, so that b^2 is even. Then b is even, and a and b have the common factor 2, contrary to the assumption that a/b is in lowest terms. ▮

Exercise 2–29. Prove that $\sqrt{3}$ is not a rational number. Suppose that $(a/b)^2 = 3$ and follow the argument of Theorem 2–5.

We extended the natural numbers N to form the integers I but still had no "more" numbers than we started with. We extended I to form Q but again had only $\aleph_0$ numbers. Now in introducing the numbers of the real field we at last bring in "more" numbers, an *uncountable* infinity of them (see Appendix B).

One easy way to understand the real numbers in their relationship to the rational numbers is to use decimal notation (or equivalently any other base, such as binary notation). Recall that in decimal notation $3.000\cdots$ is called a "terminating" decimal; when we convert $\frac{1}{11}$ to decimal form by dividing 11 into 1, we obtain a "repeating decimal," $0.090909\cdots = 0.\overline{09}$.

> **Theorem 2–6.** Every rational a/b can be written as a terminating or repeating decimal. Conversely, every terminating or repeating decimal represents a rational number.

PROOF: We obtain the decimal representation of a/b by dividing a by b. For example, we find $\frac{72}{7}$.

$$
\begin{array}{r}
10.285714285714\cdots \\
\hline
7)\overline{72.000000} \\
7 \\
\hline
2 \\
0 \\
\hline
20 \\
14 \\
\hline
60 \\
56 \\
\hline
40 \\
35 \\
\hline
50 \\
49 \\
\hline
10 \\
7 \\
\hline
30 \\
28 \\
\hline
20 \\
\end{array}
$$

$$\frac{72}{7} = 10.\overline{285714}$$

The long division problem either ends (terminating decimal) or repeats, for there are only seven possible remainders (more generally, b), 0, 1, 2, 3, 4, 5, and 6 (or $0, 1, \ldots, b-1$). If we obtained a remainder of 9, say, we would decide our trial quotient was too small and increase it by 1, leaving the correct remainder of 2. Then we can be sure of repeating a remainder and so repeating a sequence of quotient digits indefinitely. If at some point the division comes out even, the remainder is zero and the decimal terminates.

Now consider a terminating decimal, such as 6.138. From the position values of decimal notation, this can be written $\frac{6138}{1000}$, which is a rational number (not in lowest terms). In order to write a repeating decimal, d, as a rational, a/b, let $a = (10^k - 1)d$ and $b = 10^k - 1$, where k is the number of digits in the repeated part. For example, if $d = 10.\overline{285714}$, then

$$10^6 d = 10285714.\overline{285714}$$

$$d = 10.\overline{285714}$$

subtracting, $$(10^6 - 1)d = 10285704.\overline{0},$$

so that $d = \frac{10285704}{999999}$, a rational number. ▮

Exercise 2–30. Let $k_w k_{w-1} \cdots k_1 . a_1 a_2 \cdots a_f$ be a terminating decimal with w digits to the left of the decimal point, f digits to the right. Write the number in fraction form.

Exercise 2–31. Let $0.0 \cdots 0\underbrace{a_1 a_2 \cdots a_f}_{g0's,}$, $a_1 \neq 0$, be a repeating decimal. Write the number in fraction form.

If terminating and repeating decimals represent rational numbers, what do all the other decimals represent? We extend the system Q by adjoining all other decimals and call the result the real numbers R. Addition and multiplication operations can be defined so that R is a field.

> **Definition 2–16.** The **field R of real numbers** is made up of all numbers than can be represented in decimal form (including negatives and zero) with operations $+$ and $\cdot$, as will be discussed.
>
> To avoid ambiguity, we can use the convention that the non-terminating form is chosen for each decimal, choosing $2.36\overline{9}$ for 2.37, for instance. With this convention each real number has exactly one decimal form and we can say that two real numbers are equal if and only if they agree in every decimal place.

It can be shown, as in Appendix B, that the real numbers R are not countable.

To define the sum and the product of two real numbers $k_w k_{w-1} \cdots k_1 . a_1 a_2 \cdots$ and $m_v m_{v-1} \cdots m_1 . b_1 b_2 \cdots$, each of which may have indefinite length, we compute partial sums and products, arbitrarily cutting off the numbers at some decimal place, then show that as we include more and more of the decimal places of the two numbers, the partial sums and partial products tend toward limiting values which we then define to be the sum and product. The limit idea, basic to calculus and analysis, makes concrete our general idea that as we go farther and farther out in the decimal representations of the two numbers, we eventually see little appreciable change in the partial sums and partial products, in fact, less change than some small preassigned value ε. We

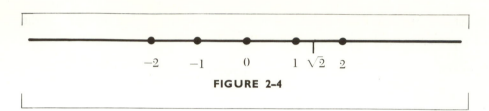

FIGURE 2-4

shall not go through proofs here that the sum limit and the product limit exist. The inverse of a real number also involves a limit process, and we shall merely note that it is possible to prove that all the required limits exist and have the desired properties, *i.e.*, the inverse property, associativity, commutativity, distributivity.

If we think of real numbers as lengths measured along a line with a specified 0, unit distance, and positive direction, we have a one-to-one correspondence between real numbers and vectors from 0 on the line. The number of points on a line (or a ray or a line segment) is $2^{\aleph_0}$ (see Appendix B). The integers are represented by the marked intervals at multiples of the unit length from 0. The rational numbers represent lengths a/b that can be divided into a equal parts, each equal to $1/b$ of the specified unit. Ancient mathematicians called a and b "commensurate," if they have a common measure: the a length can be finely enough divided to be measured in the same units, u, as b when b is finely divided: $a = ju$, $b = ku$, j and k integers.

5. The complex numbers C. In Section 2 we used equivalence classes of number pairs from the natural numbers N to introduce the integers I. Our object was to extend the natural numbers to include solutions of $x + b = a$. Then in Section 3 we used equivalence classes of number pairs from I to introduce the rational numbers Q. Our object was to extend I to include solutions of $bx = a$. Here we make use of the same pattern—equivalence classes of number pairs from the real numbers R—to form C, the complex numbers. Our object is to extend R to include solutions of quadratic equations $ax^2 + bx + c = 0$, where the coefficients a, b, and c are in R.

The reason we cannot solve every quadratic equation in R is that there is no real number r whose square is negative. One of the quadratic equations

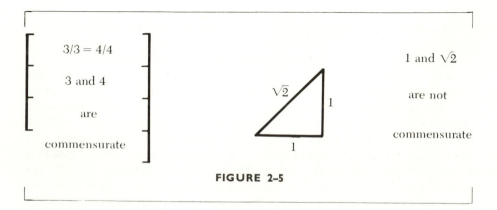

FIGURE 2-5

having no solution in R is

$$x^2 + 1 = 0.$$

If we try a solution $x = 0$, we obtain a positive left number, 1, which cannot equal 0. A positive value for x has a positive square, making the left member greater than 1. A negative value for x also has a positive square, so that the left member is greater than 1. What we need for a solution is a number whose square is -1.

It turns out that if we had such a number we could find solutions for all the quadratic equations except the degenerate case $a = 0, b = 0$. We can show this by "completing the square" in the quadratic equation, $ax^2 + bx + c = 0$. First, we want to divide both members of the equation by the coefficient a of x^2, and in doing so we come upon our first condition, that $a \neq 0$. What happens if $a = 0$? If $a = 0$, the quadratic becomes $bx + c = 0$, which we recognize as a linear equation. If $b \neq 0$, the equation has a rational solution if b and c are rational and a real solution if b and c are real. If $b = 0$, the equation has no solution. Then, supposing $a \neq 0$ so that it has an inverse $1/a$ in R, we divide both members of the original quadratic equation by a, or, which is exactly the same thing, multiply by $1/a$:

$$x^2 + \frac{b}{a} x + \frac{c}{a} = 0.$$

Now we recognize the general form of the square of a linear polynomial $x + d$, the square of which is $x^2 + 2dx + d^2$. We see that the form of our quadratic expression requires $2d = b/a$, or $d = b/2a$, so we arbitrarily introduce $d^2 = b^2/4a^2$ by adding it to both members of the equation. We also add $-c/a$ to both members:

$$x^2 + \frac{b}{a} x + \frac{b^2}{4a^2} = \frac{-c}{a} + \frac{b^2}{4a^2}.$$

Now the left member is indeed the square of $x + d$, or $x + b/2a$:

$$\left(\frac{x + b}{2a}\right)^2 = \frac{b^2}{4a^2} - \frac{c}{a} = \frac{b^2 - 4ac}{4a^2}.$$

Then we need a solution value x for which

$$\frac{x + b}{2a} = \frac{\pm\sqrt{b^2 - 4ac}}{2a}.$$

This formula is our first example of solution by radicals, in this case square roots.

If $b^2 - 4ac$ happens to be 0, then 0 serves as a square root, and we can solve $x + b/2a = 0$. If $b^2 - 4ac$ happens to be positive, there are a real positive number f and its negative $-f$ having the square $f^2 = b^2 - 4ac$. Then we can solve $x + b/2a = +f/2a$ and $x + b/2a = -f/2a$. If $b^2 - 4ac$ is negative, then $(-1)(b^2 - 4ac)$ is positive. If our system included a number whose square was -1, say i, then the products fi and $-fi$ would have the desired square.

Again we simply "play 'em as if we had 'em." Recognizing that we are outside the real number field, we take the new name "complex" numbers for the real number field with i adjoined, where $i^2 = -1$.

Later we shall develop a technical definition of field adjunction, but for now note that we need more than simply i in addition to all the real numbers if we are to have solutions for all quadratic equations. Each solution will have the form

$$x = \frac{-b \pm \sqrt{b^2 - 4ac}}{2a}.$$

The square root term gives rise to a real number or a real number times i. Since the quotient of the real numbers (denominator $2a \neq 0$) is a real number, the most general form for x is

$$x = s + ti,$$

where s and t are real. In case $b^2 - 4ac$ is nonnegative, t equals 0 and x is real.

For these reasons we consider number pairs (s, t), defining their addition and multiplication consistent with arithmetic in R when t is 0 and preserving the associative and distributive laws.

Definition 2–17. The **complex numbers C** are number pairs (s, t), where $s, t \in R$ (alternatively, forms $s + ti$, where $i^2 = -1$). Two complex numbers (s, t) and (u, v) are equal if and only if s equals u (their real parts are equal) and t equals v (their pure imaginary parts ti and vi are equal). **Addition** is defined by $(s, t) + (u, v) = (s + u, t + v)$, and **multiplication** by $(s, t) \cdot (u, v) = (su - tv, sv + tu)$. We let i stand for the pair $(0, 1)$, noting that $(0, 1)^2 = (-1, 0)$.

Theorem 2–7. The complex numbers with $+$ and $\cdot$, as defined in Definition 2–17, constitute a field.

PROOF: Directly from the definition we can see that addition and multiplication are operations on C.

Exercise 2–32. What properties of arithmetic in R do you need to prove the associative law of addition in C?

Exercise 2–33. What properties of arithmetic in R do you need to prove the associative law of multiplication in C?

Exercise 2-34. Write the additive and multiplicative identities of C and show how to obtain the additive inverse of any complex number (s, t) and the multiplicative inverse of any nonzero complex number. (For the multiplicative inverse, try multiplying (s, t) by $(s, -t)$.)

Exercise 2-35. Determine what is still necessary to prove Theorem 2-7 and complete the proof. ∎

Unfortunately, numbers involving $\sqrt{-1}$ have been stuck with the name "imaginary." In a complex number $s + ti$, s is called the "real" part and ti the "pure imaginary" part. The idea was that -1 did not "really" have a square root, but we might suppose it had an "imaginary" one. Looking at it from a modern point of view, as in our development here, i is no more imaginary than -4 or $\frac{2}{3}$, but merely convenient. If you need to think of it as a convenient fiction, that does not matter in mathematics and is typical of algebra. After all, each of us once held the point of view that "you can't subtract 10 from 6" and "you can't divide 3 into 2."

R is isomorphic to the subset of C with zero imaginary part. The addition and multiplication defined on C reduce to those defined on R in case the imaginary parts are zero.

Exercise 2-36. Verify the preceding sentence.

The number of elements in C is $2^{\aleph_0}$, the same as the number of elements in R, for they can be put into one-to-one correspondence (see Appendix C).

Comparisons Among Abstract Algebras

In a diagram like that in Figure 2-6 we can sum up the interrelations among the kinds of abstract algebras we have defined in this chapter. If in the diagram an algebra-type A is below a type B, we can say "All abstract algebras of type A are also of type B." For instance, all integral domains are groups, all groups are semigroups, and so forth. It is *not* the case that all rings with unit element are commutative.

Definition 2-18. An **abstract algebra** is a set together with various operations defined on it.

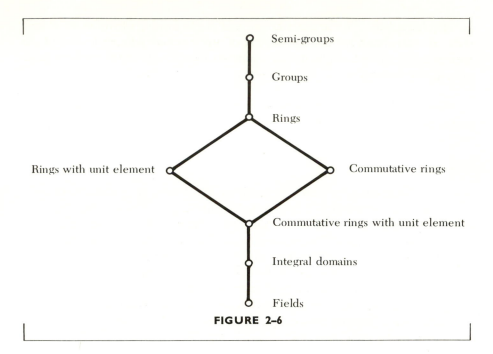

FIGURE 2–6

This definition finally justifies our chapter title and gives us a generic term for the kind of mathematical entities we have been studying. There are mathematicians who study "category theory," the theory of abstract algebras as a generic whole, and Definition 2–18 is taken from their work. Other mathematicians study a certain class of abstract algebras called "algebras," with special defining properties.

We can even carry our abstraction one step further by using it to analyze itself (see Appendix E).

To explain abstraction, consider an analogy in spelling: Our English words are spelled semiphonetically using only the 26 letters of our alphabet. With some exceptions we "abstract" the sound "B" from its many different occurrences, using the same letter to suggest the B sound whether it occurs at the beginning of a word or in the middle, whether it occurs in a long word or in a short word, whether it occurs in an adjective or in a noun. In this sense our letters can be considered abstractions.

Did you know that the counting numbers are abstractions? The concept "3" is abstracted from all our experiences with 3-membered sets, like 3 stones, 3 people, 3 blocks. If it were not such a successful abstraction, we might have to learn two separate rules for counting 3 people-plus-4-more-people or 3 stones-plus-4-more-stones! When we study "word" problems, we are simply trying to avail ourselves of the advantages of abstract arithmetic.

What are the advantages of abstraction? First there is the advantage of universality. As we remove more and more individual restrictions, our abstractions cover more and more diverse cases. Think how universal counting is. We can use the same positive integers whether we are counting steps, fruits, people, or ideas. The abstract algebra you are learning draws much of its importance from its universality; it enables us to work many seemingly different problems at once.

A second advantage of abstraction is a little harder to appreciate but is essential to understanding the importance of mathematics in our Age of Automation. To return to the spelling analogy, our reduction of words to combinations of 26 letters makes our written language feasible. The written language in turn makes it possible to analyze the spoken language, through grammar terms, descriptions of dialects, and so forth. The abstraction of mathematics enables us to analyze reasoning itself, to do a motion-study of our own thought processes.

Sometimes two systems that are isomorphic are said to be "abstractly identical." We can use this phrase to shed further light on the word "abstract" as used in "abstract algebra." If two groups are "abstractly identical," they are alike where their group properties are concerned, although one may be a group of letters or numbers, the other a group of physical rotations. That is, they are alike as *abstract algebras* only. We have already seen some widely different examples of abstract algebras, showing the universality of the concepts. In the next chapter the abstractions we have introduced concerning groups make it possible for us to analyze groups, making use of the power of analysis the abstraction provides.

Fields, especially Q, R, and C, will be important to us in studying solution of equations. As we showed in Theorem 2–5, even a quadratic equation can fail to have solutions in a certain field (Q). In the case of the equation $x^2 - 2 = 0$ we were able to find another field (R), containing a subfield isomorphic to the coefficient field Q in which the equation does have roots $+\sqrt{2}$, $-\sqrt{2}$. In Chapter 7 we shall develop a generalization of the number-pair process by which we derived I from N, Q from I, and so forth. The generalization will enable us to extend a coefficient field to include roots of equations of high degrees.

Classroom Exercise 2–37. Participate in a "spelldown" of definitions from Chapters 1 and 2. This continues the "language-study" approach to learning mathematics through oral drill and focus on vocabulary.

Classroom Exercise 2–38. Leaving out all detail, outline the successive enlargement of N to C. N stands for? Using number pairs from N we formed—?, and so forth.

Classroom Exercise 2–39.   Give reasons for these "trivial" parts of definitions:

1. A field must have at least two elements.

2. A rational number a/b cannot have a zero denominator.

3. $\frac{3}{1}$ is in Q_1, but 3 is in I.

4. In the field R we sometimes think of 3 as $2.999\cdots$. What possible advantage can there be to such a complicated form for 3?

5. Why do we assume $a \neq 0$ in $ax^2 + bx + c = 0$?

Classroom Exercise 2–40.   Learn prototypes for the abstract algebras you have studied, a familiar sample of each one.

CHAPTER 3

MORE ABOUT GROUPS

In the previous chapter we learned how groups fit into the picture of abstract algebras generally. They play such an important role that most of the algebras we defined are groups with further requirements.

In this chapter we shall develop a way to take a group apart and look at its inner structure.

Examples have shown us how a group can be represented by its "multiplication" table [more generally, its "operation" table, but we often use the familiar symbols of multiplication whatever the group operation is: a^{-1} for the inverse, ab for $a \, \theta \, b$, and 1 for the group identity]. The row headings g_i and the column headings g_j in the table are the elements of the group in the same order, the identity of the group listed first. The entry in the ith row, jth column (vertical) of the body of the table is $g_i g_j$. Then, the first row and the first column repeat the row and column headings, since they represent multiplication by the identity. A multiplication table is sometimes called a Cayley table (after the English mathematician, Arthur Cayley, 1821–1895).

·	1	i	−1	−i
1	1	i	−1	−i
i	i	−1	−i	1
−1	−1	−i	1	i
−i	−i	1	i	−1

FIGURE 3–1. Multiplication of the 4th Roots of Unity.

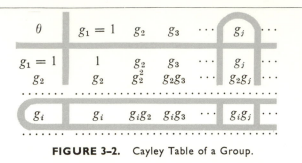

FIGURE 3–2. Cayley Table of a Group.

Theorem 3–1. The multiplication table of a group is a latin square; that is, each group element occurs once and only once in each row of the body of the table and in each column of the body of the table.

PROOF: What are the elements of the ith row? They are

$$g_ig_1, g_ig_2, g_ig_3, \ldots, g_ig_j, \ldots \tag{1}$$

Are any two of these equal? Suppose g_ig_2 and g_ig_3 were equal. Because G is a group, g_i has an inverse g_i^{-1}. If

$$g_ig_2 = g_ig_3,$$

then

$$g_i^{-1}(g_ig_2) = g_i^{-1}(g_ig_3),$$

so that

$$(g_i^{-1}g_i)g_2 = (g_i^{-1}g_i)g_3$$

by the associative law, so that

$$1g_2 = 1g_3,$$

and so

$$g_2 = g_3.$$

But $g_2 \neq g_3$. They are distinct group members, so no two of the entries (1) are equal.

Does every element of G appear in (1)? Pick an element g and see where it appears in (1). Because G is a group, G contains the inverse g_i^{-1} and the product $g_p = g_i^{-1}g$. Somewhere in the list (1), then, is the product

$$g_ig_p = g_i(g_i^{-1}g) = (g_ig_i^{-1})g = g,$$

so each element g does appear once and only once in (1).

Exercise 3–1.   Make a similar list of the entries in the jth (vertical) column. Prove that no two of them can be equal.

We need to justify an implication of uniqueness in Definition 3–2: "... *the* group of all permutations on *S* is called *the* symmetric group ..." Notice that no matter what *n* objects constitute the set *S*, whether they are letters, numerals, formulas, or people's names, we can let $\{s_1, s_2, \ldots, s_n\}$ or even $\{1, 2, \ldots, n\}$ stand for them for our present purposes. Any differences would be only in notation.

Exercise 3–8.   Let

$$\begin{pmatrix} 1 & 2 & 3 & 4 & 5 \\ 2 & 3 & 1 & 4 & 5 \end{pmatrix}$$

be a permutation *P*. What is the image of 3 under the permutation? Which elements are fixed under the permutation? Which are moved? Write the inverse permutation with respect to composition.

Exercise 3–9.   Let

$$Q = \begin{pmatrix} 1 & 2 & 3 & 4 & 5 \\ 1 & 4 & 3 & 5 & 2 \end{pmatrix}.$$

Find $P \circ Q$ and $Q \circ P$. Find Q^{-1}.

Exercise 3–10.   Find $Q^{-1} \circ P^{-1}$. Find $(P \circ Q) \circ (Q^{-1} \circ P^{-1})$.

Classroom Exercise 3–11.   A simple teaching machine is arranged so that five objects appear pictured in a row to be matched with five words in a second row. The student connects an electric wire from each picture to the appropriate word, and a light turns on if all choices are correct. How many possible wirings are there for the machine? Are any of these undesirable for the purpose? What chance is there that a random choice of matchings will be correct and so turn on the light?

There is less writing to do if we replace the permutation notation by cycle notation. Here is an example of the technique. With

$$P = \begin{pmatrix} 1 & 2 & 3 & 4 & 5 \\ 2 & 3 & 1 & 4 & 5 \end{pmatrix},$$

first write $(1$, then follow with the image $1^P = 2$:

$$(12.$$

Next, notice that *P* maps 2 into 3, so follow the 2 by 3:

$$(123.$$

Then because $3^P = 1$, we consider the first cycle closed:

$$(123).$$

Exercise 3–2.   Continuing Exercise 3–1, prove that every group element *g* appears in the *j*th column.   ■

Exercise 3–3.   Review the multiplication table of the four-group and that of the octic group. Verify that Theorem 3–1 holds in both cases.

Permutations

Definition 3–1.   A **permutation** *P* of a set *S* is a one-to-one mapping of *S* onto itself.

This means that each member *s* of *S* is mapped by *P* to exactly one member s^P of *S*, and each member of *S* appears exactly once as the image s^P of some *s*. It is handy to describe the permutation in a form we have already used in examples:

$$P = \begin{pmatrix} \cdots & s & \cdots \\ \cdots & s^P & \cdots \end{pmatrix}.$$

For each element *s* in the first row, the image s^P of *s* under the permutation *P* appears just below it in the second row.

The permutation

$$\begin{pmatrix} 1 & 2 & 3 & 4 \\ 4 & 2 & 1 & 3 \end{pmatrix},$$

for instance, maps element 1 to element 4, 2 to 2, 3 to 1, and 4 to 3. It can also be written

$$\begin{pmatrix} 4 & 3 & 1 & 2 \\ 3 & 1 & 4 & 2 \end{pmatrix}$$

or

$$\begin{pmatrix} 3 & 2 & 1 & 4 \\ 1 & 2 & 4 & 3 \end{pmatrix}$$

and so forth, though we often arbitrarily arrange the first row in counting order. Notice that the four numerals may be just a convenient code for non-numerical objects, as illustrated in the following exercises.

Exercise 3–4.   Write the $4 \cdot 3 \cdot 2 \cdot 1 = 24$ possible arrangements of all four letters *A*, *E*, *P*, and *T*. Check which permutations happen to form English words. What is the probability of making an English word from these four letters by chance arrangement? By such a probability, we mean the number of "successful" or word-making arrangements divided by the total number of possible arrangements; in this case, ＿＿ chances out of 24.

Exercise 3–5. Referring to Figure 3–1, notice that the four rows of the Cayley table represent four of the permutations of 1, i, −1, −i. Write two other permutations that are not represented. Using Theorem 3–1, show why they cannot also be rows of the Cayley table.

Exercise 3–6. Mr. (1) and Mrs. (2) Fuller and Mr. (3) and Mrs. (4) Sloat are to be seated along one side of a speaker's table. Determine by reference to Exercise 3–4 how many seating arrangements are possible for the four people. Write down (code 1, 2, 3, 4) the seating arrangements that alternate men and women. What is the probability that a random permutation of the four people will alternate men and women?

Theorem 3–2. The permutations of a set S form a group under composition.

PROOF: Composition is a binary operation on the permutations of S, for let P and Q be any two permutations: The composite $P \circ Q$ maps any element s of S into $(s^Q)^P$. (Follow Figure 3–3 for a numerical example.) Because Q is a permutation, there is exactly one image s^Q in S, which then has exactly one

Let P be

$$\begin{pmatrix} 1 & 2 & 3 & 4 \\ 2 & 3 & 4 & 1 \end{pmatrix}$$

and Q be

$$\begin{pmatrix} 1 & 2 & 3 & 4 \\ 4 & 2 & 1 & 3 \end{pmatrix}.$$

The composite mapping $P \circ Q$ maps any element into exactly one element of S. For instance, $P \circ Q$ maps 4, via 3, to 4.

Conversely, each element appears as the image of exactly one element under $P \circ Q$. For instance, there is exactly one element u for which $(P \circ Q)u = 2$. If $t = 1$, $Pt = 2$. Then choose $u = 3$, because $Q3 = 1$. $(P \circ Q)3 = 2$.

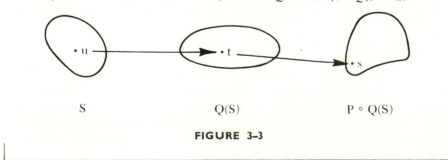

| S | $Q(S)$ | $P \circ Q(S)$ |

FIGURE 3–3

image $(s^Q)^P$ in S, because P is a permutation. Conversely, every element appears exactly once as an image $(s^Q)^P$ under the composition $P \circ Q$, for pick any s in S: There is exactly one element t whose image under P is s because P is a permutation. Then there is exactly one element u whose image under Q is t. Then $u^Q = t$, and $(u^Q)^P = P \circ Q(u) = s$.

Because permutations are mappings of a set onto itself, we can appeal to Theorem 1–3 to prove that they are associative under composition.

The identity map

$$P_1 : \begin{pmatrix} \cdots & s & \cdots \\ \cdots & s & \cdots \end{pmatrix}$$

forms the group identity. The inverse of a permutation that maps s into s^P is a permutation that maps s^P into s, so the inverse of a permutation can be written by interchanging the two rows:

$$P^{-1} = \begin{pmatrix} \cdots & s & \cdots \\ \cdots & s^P & \cdots \end{pmatrix}^{-1} = \begin{pmatrix} \cdots & s^P & \cdots \\ \cdots & s & \cdots \end{pmatrix}.$$

Definition 3–2. If S is a finite set of n elements, then the group of all permutations on S is called the **symmetric group S_n on n elements.** S_n has $n!$ members.

Let us check the order of, or number of members in, S_n. Since S has a finite number of elements, we can list them in some order as

$$\begin{pmatrix} s_1 & s_2 & s_3 & \cdots & s_n \\ \cdot & \cdot & \cdot & \cdots & \cdot \end{pmatrix}.$$

How many different second rows, and therefore how many different permutations, can we form? A permutation can take s_1, say, into any of the n elements of S. (Can s_1 be mapped to s_1? Consider, for instance, the identity permutation.) Having chosen an image t_1 for s_1 from S, we have only $n - 1$ possible images t_2 for s_2, because no two elements have the same image under a permutation, so $t_2 \neq t_1$. Then with two elements used already as images, there are only $n - 2$ choices for t_3, and so forth. Then the number of different permutations is

$$n(n - 1)(n - 2) \cdots (n - n + 1) = n(n - 1)(n - 2) \cdots 1 = n!$$

Exercises like the following one show why the numbers of choices are multiplied.

Exercise 3–7. Write the $3! = 3 \cdot 2 \cdot 1 = 6$ permutations on the set $\{1, 2, 3\}$.

Since 4 maps to itself, we have a 1-cycle (4), which can be omitted or included optionally. (5) can be omitted also. We write P as a product of cycles:

$$P = (123)(4)(5) = (123).$$

In general, to express a permutation of n elements as a product of cycles, start the first cycle with the first element s_1. Continue to follow each element of the cycle with its image under the permutation until the image s_1 appears, closing the first cycle. If the first cycle has not exhausted the n elements, start a new cycle with an element that has not been used. Continue this process until all elements have been used. Cycles of one element, or 1-cycles, may be dropped.

We can reinterpret a cycle as a permutation, except that we might not know of some dropped 1-cycles. For example, as a permutation on 6 elements, $(13645) = \begin{pmatrix} 1 & 2 & 3 & 4 & 5 & 6 \\ 3 & 2 & 6 & 5 & 1 & 4 \end{pmatrix}$, for the image of 1 is shown just after "1" in the cycle as 3, 2 is not affected by the cycle since it does not appear, the image of 3 is 6, the image of 4 is 5, and the image of 5 "follows" it in the cyclical sense, as the initial "1". The image of 6 is 4.

To multiply cycles, we keep in mind the permutations they denote, applying the right cycle first. For example, $(1234)(2465) = \begin{pmatrix} 1 & 2 & 3 & 4 & 5 & 6 \\ 2 & 1 & 4 & 6 & 3 & 5 \end{pmatrix}$, for 1 is left unchanged by the first, or right, cycle, then mapped to 2 by the left cycle; 2 is mapped to 4 by the right cycle, which is then mapped to 1 by the left cycle; 3 is unchanged by the right cycle, then mapped to 4 by the left cycle; 4 is mapped to 6 by the right cycle, and then 6 is unchanged by the left cycle; and so on.

Exercise 3–12. Write the Q of Exercise 3–9 as a cycle, (245). Find $P \circ Q$ in cycle form and compare with the permutation form.

Exercise 3–13. Write the permutations of S_3 as cycles.

Exercise 3–14. Write in permutation form $\begin{pmatrix} 1 & 2 & 3 & 4 \\ . & . & . & . \end{pmatrix}$:

(134)

(12)(34)

(1423)

(13)(12)

(14)(13)(12)

Definition 3–3. A c-**cycle** $(s_1 s_2 \cdots s_c)$ is a permutation of c different elements of a set S, mapping s_i into s_{i+1}, for $i = 1, 2, \ldots, c - 1$, and s_c into s_1. We represent composition of two cycles by writing them in adjacent positions, with the right mapping performed first. A 2-cycle is called a **transposition.**

Theorem 3–3. Every c-cycle, $c > 1$, can be written as a product of transpositions. Every permutation of a finite number $n > 1$ of elements can be expressed as a product of transpositions.

PROOF: Follow the special case in Figure 3–4 to clarify the notation of the general proof.

The product of transpositions

$$(s_1 s_c)(s_1 s_{c-1}) \cdots (s_1 s_3)(s_1 s_2) \tag{2}$$

equals the cycle $(s_1 s_2 s_3 \cdots s_{c-1} s_c)$, for the product (2) performs the same mapping as the cycle: In (2), the transposition $(s_1 s_2)$ maps s_1 to s_2. Since s_2 does not appear in the other transpositions it remains unchanged by the others. The transposition $(s_1 s_2)$ maps s_2 to s_1. Then the transposition $(s_1 s_3)$ maps s_1 to s_3, which remains unchanged by the rest of the transpositions. In general, if $1 \leq i < c$, then s_i appears first in the transposition $(s_1 s_i)$, which maps s_i to s_1, after which the next transposition, $(s_1 s_{i+1})$, maps s_1 to s_{i+1}. The later transpositions leave s_{i+1} unchanged, so that the net effect is $s_i \to s_{i+1}$. Now, s_c appears only in the last transposition, which maps it to s_1.

A permutation P of a finite number n of elements can be factored into n or fewer cycles, for start the first cycle with the first element s_1 and follow it with its image s_1^P. Follow each element of the cycle with its image under P until one of the images is s_1. Then the minimum length of the cycle is 1 in case s_1 is fixed by P; the maximum length is n, since s_1 must appear as the image of some element. The number of elements not used in the first cycle is less than the original n. In this way the n elements must be exhausted in n or fewer cycles, and each cycle can in turn be replaced by a product of transpositions, as shown in the previous paragraph. ∎

Notice that each permutation can be written many different ways as a product of transpositions—infinitely many, allowing for repetitions: For example, take the identity permutation on five elements,

$$\begin{pmatrix} 1 & 2 & 3 & 4 & 5 \\ 1 & 2 & 3 & 4 & 5 \end{pmatrix}.$$

It can be written (12)(12) or (25)(25) or (12)(34)(12)(34) or (12)(23)(23)(12) or (45)(45)(45)(45)(45)(45)(45)(45).   Each of these factorizations has an even number of transpositions.

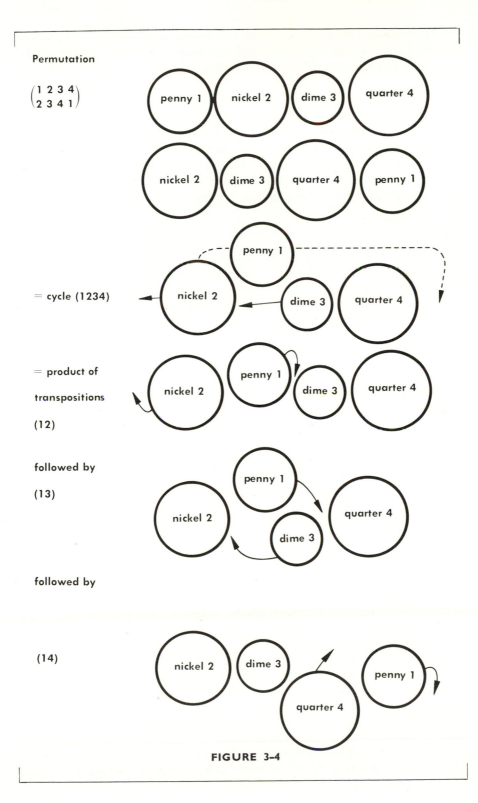

Permutation

$\begin{pmatrix} 1\ 2\ 3\ 4 \\ 2\ 3\ 4\ 1 \end{pmatrix}$

penny 1   nickel 2   dime 3   quarter 4

nickel 2   dime 3   quarter 4   penny 1

= cycle (1234)

penny 1   nickel 2   dime 3   quarter 4

= product of

transpositions

(12)

nickel 2   penny 1   dime 3   quarter 4

followed by

(13)

penny 1   nickel 2   dime 3   quarter 4

followed by

(14)

nickel 2   dime 3   quarter 4   penny 1

FIGURE 3–4

The following theorem states that the oddness or evenness of the number of transpositions (the *parity*, in mathematical language) is invariant among the many representations for any one permutation.

Theorem 3–4. If a permutation P can be written as the product of t transpositions, then every factorization of P into transpositions has $t + 2i$ transpositions, where i is an integer. (That is, either all factorizations have an even number of transpositions or all have an odd number.)

PROOF: The essential thing about a permutation on n elements is what it does to the order of those elements. One way to check on the order before and after the permutation is to write down the order of every pair of elements before the permutation and then see which orders have changed afterward. If we represent the elements by

$$1 \quad 2 \quad 3 \quad \cdot \quad \cdot \quad \cdot \quad n-1 \quad n$$

before the permutation, our comparisons of pairs tell us

$$1 < 2, 1 < 3, \ldots, 1 < n-1, 1 < n,$$

$$2 < 3, \ldots, 2 < n-1, 2 < n,$$

$$\ldots, 3 < n-1, 3 < n,$$

$$\cdot$$
$$\cdot$$
$$\cdot$$

$$n - 1 < n.$$

The symbol $<$ here can stand for "is to the left of" or "is below" or "is dependent upon" or whatever suits the situation. Similarly, the numerals $1, 2, \ldots, n$ stand not necessarily for numbers but for people or coins or playing cards or whatever is being permuted.

Now after a permutation P has been performed, we can check to see whether it changed an odd number or an even number of the $n(n-1)/2$ pair-comparisons. (The identity permutation changes none, and we take 0 to be even.)

What does a single transposition do to the pair comparisons? Take the transposition (ij) with $i < j$, meaning i is to the left of j. The transposition definitely changes the pair-comparison $i < j$ into $j < i$, for it interchanges i and j. Take any k that is to the left of both i and j. Then (ij) changes $k < i$ and $k < j$ into $k < j$ and $k < i$, so that they appear in new places in our list of comparisons, but they are unchanged. Take any k that is to the right of both i and j. Then (ij) changes $i < k$ and $j < k$ into $j < k$ and $i < k$, still unchanged. Now take any k that lies between i and j. For such a k, (ij) changes $i < k$ and $k < j$ into $j < k$ and $k < i$; that is, it changes both of them, an even number. In all other comparisons i and j fail to appear, so the net result of the transposition is to change an odd number of pair comparisons. The transposition $(ji), j > i$, performs exactly the same interchange as (ij). A product of an odd number of transpositions will have the effect of changing an odd number of pair-comparisons, an even number changing an even number, since each transposition changes an odd number (the sum of an odd number of odd numbers is odd; the sum of an even number of odd numbers is even). If a permutation P can be written as a certain product of transpositions, the oddness or evenness of their number must agree with the oddness or evenness of the number of pair-comparisons changed by P. ∎

Definition 3–4. A permutation on n elements is called an **odd permutation** if it can be factored into an odd number of transpositions: otherwise it is called an **even permutation.**

Exercise 3–15. You are going to play a game with five markers initially arranged as in Figure 3–5. The object is to beat your opponent at putting the markers in alphabetical order, left-to-right. Each turn consists of interchanging two markers, that is, performing one transposition, the winner being the one who successfully puts all five markers in order on his turn. Should you magnanimously let your opponent have the first move, or hold out for the first move yourself? Use Theorem 3–4 to analyze the strategy of this game: Which move decides who can win?

A number of board games are based on Theorem 3–4, including some of those with movers mounted in a frame allowing just one blank square. These can be mounted initially in a permutation that does not admit a solution! Such frustrating games were very popular in European drawing rooms at the time this theorem became known.

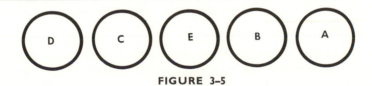

FIGURE 3–5

Permutations and their oddness or evenness are interesting in their own right, but we go on to link them even more tightly with groups. We have seen in Theorem 3–2 that all the permutations of a set S form a group. Now we shall prove that the elements of any group can be thought of as permutations.

Theorem 3–5 (Cayley's Representation Theorem for Groups). Every group G is isomorphic to a group of permutations. If $|G| = n$, then G is isomorphic to a group of permutations on n objects.

PROOF: Consider the multiplication table for G. From Theorem 3–1 each row contains every element of G exactly once. Set up permutations

$$\begin{pmatrix} g_1 = 1 & g_2 & \cdots & g_j & \cdots \\ g_i g_1 = g_i & g_i g_2 & \cdots & g_i g_j & \cdots \end{pmatrix},$$

the first row being the first row of the body of the multiplication table and the second row the ith row of the multiplication table. Set up the one to-one correspondence

$$g_i \sim \begin{pmatrix} g_1 & g_2 & \cdots & g_j & \cdots \\ g_i g_1 & g_i g_2 & \cdots & g_i g_j & \cdots \end{pmatrix}. \tag{3}$$

Now we want to show that G with its group operation $*$ is isomorphic to the permutations with their operation composition. That is, we want to show that if we do an arithmetic problem $g_i * g_k$ in G and the corresponding one,

$$\begin{pmatrix} g_1 & g_2 & \cdots & g_j & \cdots \\ g_i g_1 & g_i g_2 & \cdots & g_i g_j & \cdots \end{pmatrix} \circ \begin{pmatrix} g_1 & g_2 & \cdots & g_j & \cdots \\ g_k g_1 & g_k g_2 & \cdots & g_k g_j & \cdots \end{pmatrix},$$

the results correspond to each other according to the correspondence (3).

Working first in G, we obtain $g_i * g_k = g_i g_k$. Now evaluating the composition of the two corresponding permutations, we see that the right permutation maps each element g into $g_k g$, its left-multiple by g_k. Then the left permutation maps the result, $g_k g$ into its left-multiple by g_i, or $g_i(g_k g)$. Since G is a group, the associative law holds and gives us

$$g_i(g_k g) = (g_i g_k)g,$$

so that the composite permutation maps g into its left-multiple by $g_i g_k$. Therefore, the composite does correspond to $g_i g_k$, and the isomorphism is established. ∎

We claimed that the particular set of permutations used was a group. The fact that the permutations are isomorphic to the original group G establishes that they form a group, for the identities and inverses must all correspond. (See the next three exercises.)

Exercise 3–16. Prove that if two finite groups are isomorphic, they must have the same number of elements.

Classroom Exercise 3–17. Prove that if two groups are isomorphic, their respective identity elements correspond under the isomorphism.

Exercise 3–18. Prove that if two groups are isomorphic and if an element g of one corresponds to the element h of the other under the isomorphism, then g^{-1} corresponds to h^{-1}.

Exercise 3–19. Write the four-group as a permutation group. Write the octic group as a group of permutations expressed as cycles.

Exercise 3–20. Is the permutation group of Theorem 3–5 always a full symmetric group? (Refer to Exercise 3–5.)

Classroom Exercise 3–21. Find some permutations on four letters that are *not* included in the octic group. Interpret them as mappings of the vertices of a square, and show how each fails to be a "symmetry" of the square, that is, a distance-preserving mapping.

Classroom Exercise 3–22. Two rows of a certain Cayley table are

$$r \quad q \quad s \quad 1 \quad u \quad v \quad t \quad w$$
$$q \quad s \quad 1 \quad u \quad v \quad t \quad w \quad r.$$

The rows show the eight distinct group members. The group identity is "1." Deduce all the rest of the table and compare the group with the octic group.

Subgroups

Definition 3–5. Let G be a group with operation $*$. If a subset of the elements of G form a group H under $*$, it is called a **subgroup** of G. If H is neither the identity alone nor the whole group G, it is called a **proper subgroup** of G.

Every group G has the subgroup $\langle 1 \rangle$, with the identity of G as its only element. Notice that $\langle 1 \rangle$ is closed under $*$ because $1 * 1 = 1$, it is associative, and it has identity 1 and inverse $1^{-1} = 1$.

Every group G has itself as a subgroup, for the elements of G are a subset of the elements of G and form a group under $*$.

Notice that Definition 3–5 specifies $*$, the operation of G, as the operation of the subgroup H. Otherwise, any subset of the elements of G could be taken to be a subgroup under some suitably defined operation.

Let G be the four-group, with multiplication table

$*$	1	a	b	c
1	1	a	b	c
a	a	1	c	b
b	b	c	1	a
c	c	b	a	1

Notice that H_1: $\{1, a\}$, $*$ is a subgroup of G. First, we show that $*$ is an operation on H_1.

$$1 * 1 = 1 \in \{1, a\}$$
$$1 * a = a * 1 = a \in \{1, a\}$$
$$a * a = 1 \in \{1, a\},$$

so H_1 is closed under $*$. H_1 is associative under $*$, for each triple of elements of H_1 is also a triple of elements of G, for which the associative law is known. The identity 1 is in H_1. Also, $a^{-1} = a$ is in H_1.

Exercise 3–23. Show that H_2: $\{1, b\}$, $*$ and H_3: $\{1, c\}$, $*$ are subgroups of the four-group G.

Exercise 3–24. Show that J: $\{1, a, b\}$, $*$ is not a subgroup of G. Show that if an operation ψ is defined on $\{1, a, b\}$ by the multiplication table

ψ	1	a	b
1	1	a	b
a	a	b	1
b	b	1	a

then J, ψ is a group.

Theorem 3–6. Let G be a group with operation $*$, and let H be a nonempty subset of the elements of G. Then H is a subgroup of G if and only if

i. H is closed under $*$

and

ii. H has the inverse property; that is, for every element $h \in H$ there is an element $h^{-1} \in H$ for which $h^{-1} * h = h * h^{-1} = 1$, the identity element under $*$ in G.

If G is finite, then i alone forms a necessary and sufficient condition that H be a subgroup of G.

PROOF: To prove the necessity, or "only if" quality, of requirements *i* and *ii*, we suppose that *H* is a subgroup of *G* and prove that *i* and *ii* must then hold.

If *H* is a subgroup of *G*, then, by Definition 3–5, *H* is a group under *, so, by Definition 1–2, * is a binary operation on *H*, which, according to Definition 1–1, guarantees that *H* is closed under *. Schematically, we have

Hypothesis: *H* is a subgroup of *G*.

Deductions	*Reasons*
1. *H* is a group under *.	1. Definition 3–5 of subgroup
2. * is binary operation on *H*.	2. Definition 1–2 of group
3. *H* is closed under * (or *i*).	3. Definition 1–1 of binary operation

Still assuming that *H* is a subgroup, we see that *H* has the inverse property, because, by Definition 3–5, *H* is a group and, by Definition 1–2, groups have this property.

Now to prove that *i* and *ii* are sufficient to imply that a nonempty subset *H* of *G* is a subgroup of *G*, we assume *i* and *ii* as hypothesis and deduce that *H* satisfies Definition 3–5. First, is * a binary operation on *H*? Because * is a binary operation on *G*, it is a mapping from $G \times G$ to *G* (Definition 1–1), so it is defined on $H \times H$ also. By the hypothesis *i* of closure, it is a mapping to *H*. Therefore * maps $H \times H$ to *H* and so is a binary operation on *H*.

Next, does the associative law hold? Yes, because any triple of elements h_1, h_2, h_3 of the subset *H* is a triple of elements of the set of elements of *G*, where we know that the associative law holds.

Is the identity element 1 of *G* a member of *H*, and does it act as an identity in *H*? Yes. First, notice that *H* is nonempty by hypothesis, so let *h* be one of its members. Then from hypothesis *ii* h^{-1} is a member of *H*. Then by closure $h * h^{-1} = 1$ is also a member of *H*.

The inverse property of groups is assumed for *H* in hypothesis *ii*. This completes the proof except for the simplification when *G* is finite.

When *G* is finite, closure alone is necessary and sufficient for a nonempty subset *H* to form a subgroup of *G*. Certainly it is necessary, for *i* and *ii* have both been proved necessary without the restriction of finiteness. We establish the sufficiency of *i* alone in the finite case by showing that when *G* is finite we can deduce *ii* from *i*. Suppose *G* has a finite number *n* of elements. By hypothesis, *H* has at least one element. Let *h* be any element of *H*. Consider powers of *h*: h, h^2, h^3, ..., h^n, h^{n+1}. From hypothesis *i* of closure all these powers of *h* are elements of *H*. But because $|H| \leq |G| = n$, not all $n + 1$ powers of *h* can be different. Suppose $h^e = h^f$, where $f = e + d$, $d > 0$. Then $h^e = h^f = h^{e+d} = h^e h^d$. Since h^e is in *G*, it has an inverse. Multiplying the equation $h^e = h^e h^d$ on the left by that inverse, we have $1 = h^d$. Since 1 can be expressed as a power of *h*, closure guarantees that 1 is in *H*. If $d > 1$, $h^{d-1} \in H$ supplies the required inverse of *h*. In case $d = 1$, $h^{-1} = h = 1$. ∎

The proof of Theorem 3–6 is elementary but rather complicated; that is, we have used quite basic results, but we have used them in an involved sequence. You might contrast this proof and the skills you need to follow it with the numerical computations you once considered typical of mathematics. This proof provides a good example of what mathematicians really do.

Theorem 3–6 makes it fairly easy to check for subgroups of a group. Many of the groups we shall work with, such as the symmetric group S_n, are finite, so that we need only verify that a subset is closed under the group operation to show that it is a subgroup.

Exercise 3–25. Use Theorem 3–6 to prove that S_m is isomorphic to a subgroup of S_n if $m \leq n$. (Let the permutation of S_m correspond to the permutations of S_n that leave some of the n objects fixed.)

Exercise 3–26. Prove that the even permutations of S_n form a subgroup of S_n.

Exercise 3–27. Use Theorem 3–6 to find five subgroups (not all proper) of the four-group.

Exercise 3–28. Either prove this conjecture or disprove it with a counterexample: *Conjecture.* The identity of a subgroup H of G is necessarily the identity of the group G. Notice that among the mappings of a square treated in Chapter 1 (octic group), the identity is the mapping that does not change the square; it is also a rotation of 0° and a null reflection in each axis.

Exercise 3–29. Either prove or disprove: *Conjecture.* The inverse of an element in a subgroup H of G is the same as its inverse in G.

Congruence and Cosets

We make here a tentative definition, then we prove a theorem that points to our revised definition. We proceed in this unorthodox way to give you some of the flavor of mathematics in its "art" stage when the definitions and premises are being concocted. While postulational systems are quite arbitrary and there is no moral suasion for or against adopting any suggested definition, still we "try on" different ones to see what line their consequences will take, so as to pick those that will lead to an interesting system, preferably one in which we can prove some theorems.

Tentative Definition. Let S be a nonempty subset of the elements of a group G. Consider a relation "congruence modulo S" defined for elements of G by: g_1 is congruent to g_2 modulo S, written $g_1 \equiv g_2 \pmod{S}$ if $g_1 = g_2 s$ for some element s in S.

We use this tentative definition in the following theorem.

Theorem 3–7. Let S be a nonempty subset of the elements of the group G. Congruence modulo S is an equivalence relation if and only if S is a subgroup of G.

PROOF: First suppose congruence modulo S is an equivalence relation. According to Definition 2–13, an equivalence relation is transitive and symmetric. Then given $s, t \in S$, we have $t = 1 \cdot t$ and $1 = s^{-1} \cdot s$, since $s, t \in G$. Then since $t \equiv 1$ and $1 \equiv s^{-1}$, we have from transitivity that $t \equiv s^{-1}$. That is, there is a $u \in S$ for which $t = s^{-1}u$. Then $st = u \in S$. This proves that the subset S is closed under the group operation of G, which is sufficient to conclude that S is a subgroup of G if G is finite (Theorem 3–6).

In the more general case the group G may be infinite, and we need to prove that each element s of S has an inverse in S. Let $s \in S$. Then $1 = s^{-1} \cdot s$, because $s \in G$, so $1 \equiv s^{-1}$. Then by symmetry of the equivalence, we can write $s^{-1} \equiv 1$, which means there is a v in S for which $s^{-1} = 1 \cdot v = v$. Then s^{-1} is in S.

Exercise 3–30. Show how to prove that 1 is in S from the reflexive property of an equivalence. Why does Theorem 3–6 make this proof unnecessary in showing that S is a subgroup?

From Theorem 3–6 we now have enough to conclude that S is a subgroup of G, for it is a nonempty subset closed under the group operation and having the inverse property.

To prove the converse, suppose that S is a group. Is congruence reflexive? Yes, there is always an $s \in S$ for which $g_1 = g_1 s$, for we can let s be 1, the identity of S. Is congruence symmetric? Yes, for if $g_1 = g_2 s$, then $g_2 = g_1 s^{-1}$. Is congruence transitive? Yes, for if $g_1 = g_2 s$ and $g_2 = g_3 t$, then $g_1 = g_3(ts)$. ∎

Because of this result, we revise our definition of congruence to require the subset to be a subgroup under the operation of G, and we thus establish an equivalence relation.

Definition 3–6. Let H be a subgroup of a group G. Then two elements g_i, g_j of G are congruent modulo H if there is an element h_k in H for which $g_i = g_j h_k$. We may call this **left congruence,** defining right congruence as in Exercise 3–31.

Now suppose that H is a subgroup of a group G. Then by Theorem 3–7 congruence modulo H is an equivalence relation in G. Then congruence modulo H partitions G into mutually exclusive and exhaustive equivalence classes, each part made up of elements that are congruent to each other. (See discussion of equivalence classes following Definition 2–13, page 28.) From those elements congruent to $g_1 = 1$, we obtain

$$\begin{cases} g_1 h_1 = h_1 \\ g_1 h_2 = h_2 \\ \quad \cdots \cdots \\ g_1 h_k = h_k \\ \quad \cdots \cdots \end{cases}$$

We use $g_1 H = H$ as an abbreviation for this whole subset of G. From the elements congruent to g_2 we obtain $g_2 H$, short for

$$\begin{cases} g_2 h_1 \\ g_2 h_2 \\ \quad \cdots \\ g_2 h_k \\ \quad \cdots \end{cases},$$

and so on.

> **Definition 3–7.** Let H be a subgroup of a group G. The **left cosets of H in G** are the sets $gH = \{gh_1, gh_2, \ldots, gh_k, \ldots\}$, where $g \in G$ and the h's are in H.

Notice that $g_i \equiv g_j \pmod{H}$ if and only if their respective equivalence classes, or cosets, are equal, *i.e.*, $g_i H = g_j H$. To obtain distinct cosets, we use elements g_i, g_j that are incongruent modulo H.

> **Definition 3–8.** Let H be a subgroup of a group G. The number of distinct left cosets of H in G is called the **index of H in G,** written $[G:H]$.

Exercise 3–31. Define "right congruence" modulo a subgroup H in G, analogous to Definition 3–6, by $g_i = h_m g_j$. Prove that right congruence is an equivalence relation.

Exercise 3–32. Define "right cosets" based on right congruence (Exercise 3–31) analogous to Definition 3–7. The coset Hg need not equal the coset gH, but show that there are the same number of distinct right cosets as left cosets, so that the index is independent of which type of coset is used. (Set up a one-to-one correspondence $g_i H \leftrightarrow Hg_i^{-1}$ between the left cosets and the right cosets.)

Theorem 3–8 (Lagrange). Let a group G have order $|G|$ and its subgroup H have order $|H|$. Then $|G| = |H| \cdot [G:H]$. If G is finite, then $|H|$ divides $|G|$.

PROOF: Each coset has $|H|$ elements, and there are $[G:H]$ distinct cosets. The cosets are "mutually exclusive and exhaustive," since they are equivalence classes. The total count $|G|$ is then $|H| \cdot [G:H]$. If $|G|$ is finite, $|H|$ divides $|G|$ evenly with the quotient $[G:H]$. ∎

A few examples are necessary before we go further into the analysis of a group, its subgroups, and their cosets.

Example 1: The Four-Group. Consider the Klein four-group $\{1, a, b, c\}, \cdot$ as the group G and its subgroup $\{1, a\}, \cdot$ as H. Recall the multiplication table

	1	a	b	c
1	1	a	b	c
a	a	1	c	b
b	b	c	1	a
c	c	b	a	1

The left cosets of H in G are

$$1H = \begin{cases} 1 \cdot 1 = 1 \\ 1 \cdot a = a \end{cases} \quad bH = \begin{cases} b \cdot 1 = b \\ b \cdot a = c \end{cases}$$

We do not need to include aH and cH, since they equal cosets we have already included:

$$aH = \begin{cases} a \cdot 1 = a \\ a \cdot a = 1 \end{cases} = 1H \quad cH = \begin{cases} c \cdot 1 = c \\ c \cdot a = b \end{cases} = bH$$

$|G| = 4$, $|H| = 2$, and $[G:H] = 2$. $4 = 2 \cdot 2$, illustrating Lagrange's Theorem 3–8.

Example 2: The Octic Group or Mappings of a Square. In Chapter 1 we described eight symmetries, or rigid mappings, of a square. They can be written in cycle form in terms of how they map the vertices:

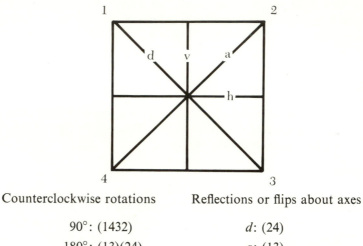

Counterclockwise rotations	Reflections or flips about axes
90°: (1432)	d: (24)
180°: (13)(24)	a: (13)
270°: (1234)	v: (12)(34)
Identity map: (1)	h: (14)(23)

The multiplication table under composition for these mappings is pictured in Figure 3–6.

The rotations together with the identity mapping form the subgroup $R = \{1, 90, 180, 270\}$, ∘. (We drop the degree symbol at this point.) The left cosets of R in the octic group G are

$$1 \circ R = \begin{cases} 1 \circ 1 = 1 \\ 1 \circ 90 = 90 \\ 1 \circ 180 = 180 \\ 1 \circ 270 = 270 \end{cases} \quad \text{and} \quad d \circ R = \begin{cases} d \circ 1 = d \\ d \circ 90 = v \\ d \circ 180 = a \\ d \circ 270 = h \end{cases}$$

$$|G| = 8; \quad |R| = 4; \quad [G:R] = 2. \quad 8 = 4 \cdot 2.$$

∘	1	90	180	270	d	a	v	h
1	1	90	180	270	d	a	v	h
90	90	180	270	1	h	v	d	a
180	180	270	1	90	a	d	h	v
270	270	1	90	180	v	h	a	d
d	d	v	a	h	1	180	90	270
a	a	h	d	v	180	1	270	90
v	v	a	h	d	270	90	1	180
h	h	d	v	a	90	270	180	1

FIGURE 3–6. Multiplication table for the octic group.

Now we find the left cosets in G of the subgroup $V = \{1, v\}, \circ$.

$$1 \circ V = \begin{Bmatrix} 1 \\ v \end{Bmatrix}; \quad 90 \circ V = \begin{Bmatrix} 90 \\ d \end{Bmatrix}; \quad 180 \circ V = \begin{Bmatrix} 180 \\ h \end{Bmatrix}; \quad 270 \circ V = \begin{Bmatrix} 270 \\ a \end{Bmatrix}.$$

$$|G| = 8, |V| = 2, [G:V] = 4.$$

Compare these with the *right* cosets of V in G.

$$V \circ 1 = \begin{Bmatrix} 1 \\ v \end{Bmatrix}; \quad V \circ 90 = \begin{Bmatrix} 90 \\ a \end{Bmatrix}; \quad V \circ 180 = \begin{Bmatrix} 180 \\ h \end{Bmatrix}; \quad V \circ 270 = \begin{Bmatrix} 270 \\ d \end{Bmatrix}.$$

We find that $1 \circ V = V \circ 1$ and $180 \circ V = V \circ 180$, but that the left coset $90 \circ V$ does not equal the right coset $V \circ 90$. In fact, $90 \circ V = V \circ 270$ and $270 \circ V = V \circ 90$.

This noncommutative group, then, shows us an example of an individual left coset that fails to equal its corresponding right coset, but, as proved in Exercise 3–32, the number of different cosets $[G:H]$ remains invariant whether we use left cosets or right cosets. Note that 270 is the inverse of 90, illustrating the correspondence suggested in Exercise 3–32 between $90 \circ V$ and $V \circ (90)^{-1}$.

Exercise 3–33. Find the left cosets of D in G, where $D = \{1, d\}, \circ$ and G is the octic group. Find the right cosets. Show that $|G| = |D| [G:D]$ in both cases.

Exercise 3–34. Show that by combining the two mappings 90 and v, using each as many times as you like, you can generate the whole octic group. This means that any subgroup containing these two mappings must equal the whole octic group. Why?

Example 3: *Cyclic Groups.* We shall see that each element of a finite group generates a subgroup.

> **Theorem 3–9.** Let g be an element of a finite group G of order n. Then there is a minimal positive exponent e for which $g^e = 1$ and e divides (evenly) any positive exponent f for which $g^f = 1$. The distinct powers of g form a subgroup $\langle g \rangle$ of G. Also, e divides n.

PROOF: Not all the first $n + 1$ powers of g,

$$g^1, g^2, g^3, \cdots, g^n, g^{n+1},$$

$$g, g^2, g^3, g^4, g^5, g^6, g^7, g^8, g^9, g^{10}, g^{11}, g^{12}, g^{13}, \ldots$$

Suppose $g^7 = g^{11}$. Then $g^7 = g^4 g^7$, so that

$$g^7(g^7)^{-1} = g^4 g^7 (g^7)^{-1}, \quad \text{or} \quad 1 = g^4.$$

The listing of the 13 powers of g can then be replaced by

$$g, g^2, g^3, 1, g, g^2, g^3, 1, g, g^2, g^3, 1, g, \ldots$$

FIGURE 3–7

are distinct, for they are all members of G, which has only n members. (Follow Figure 3–7 for a special case.) Then suppose $g^s = g^t$ with $s < t$. We multiply

$$g^s = g^t = g^{t-s} g^s$$

by $(g^s)^{-1}$, obtaining

$$1 = g^{t-s},$$

with $0 < t - s$. This shows that there are positive exponents $t - s$ for which $g^{t-s} = 1$. Let e be the least such positive exponent. Then the powers of g are periodic, with distinct powers equal to

$$g, g^2, g^3, \ldots, g^{e-1}, g^e = 1,$$

since $g^{e+1} = g^e g = 1g = g$, $g^{e+2} = g^e g^2 = 1g^2 = g^2$, and so forth and since $g^s = g^t$ with $1 \le s < t < e$ would give us $t - s < e$.

Suppose f is a positive exponent for which $g^f = 1$. Then since e is the least such positive exponent, $f \ge e$. We divide f by e, obtaining a quotient q and remainder $r < e$,

$$f = eq + r.$$

Then $g^f = g^{eq+r} = (g^e)^q g^r = 1^q g^r = g^r = 1$. But $r < e$, so r must be zero since g^e is the least positive power of g that equals 1. Thus $f = eq$, and e divides f.

By Theorem 3–6 the distinct powers of g form a subgroup of G, for they are closed under the group operation: $g^x g^y = g^{x+y}$, which can be reduced by powers of g^e. Then by Lagrange's Theorem 3–8, e divides n, the quotient being the index of the subgroup in G. ∎

Definition 3–9. A group $\langle g \rangle$ made up entirely of powers of an element g is called a **cyclic group** with g as a **generator.** If $\langle g \rangle$ is finite, the least positive e for which $g^e = 1$ is called the **order of g,** and the distinct group elements are $g, g^1, g^3, \ldots, g^e = 1$. If $\langle g \rangle$ is an **infinite cyclic group,** then $g^s = g^t$ only if $s = t$. An example of an infinite cyclic group is that generated by the permutation of integers

$$g = \begin{pmatrix} \cdots & -3 & -2 & -1 & 0 & 1 & 2 & 3 & \cdots \\ \cdots & -2 & -1 & 0 & 1 & 2 & 3 & 4 & \cdots \end{pmatrix}.$$

A cyclic group is necessarily abelian (see Exercise 2–7).

Exercise 3–35. In trigonometry the function sine A is periodic. Graph one full period of the function. Trigonometry tables do not list sine 400°. How can you find its equivalent in the tables?

Exercise 3–36. Consulting Example 2 on the octic group, show that R is a cyclic subgroup. Write out a complete cycle of the mapping 90. What is the order of the mapping? Letting $(90)^{379}$ represent the composition of the 90° mapping taken 379 times, find its equivalent mapping in the octic group.

Exercise 3–37. In the octic group find the cyclic subgroup generated by the 270 mapping. What is the order of this mapping? What is the index of the subgroup in the octic group?

Exercise 3–38. In the octic group show that $270 \circ 90 = 90 \circ 270 = 1$, so that $270 = (90)^{-1}$. Write each of the members of the cyclic subgroup in Exercise 3–37 as a power of 90.

Exercise 3–39. Prove that a group element and its inverse have the same order.

Exercise 3–40. Write the multiplication table for the cyclic group of size four, with elements $g, g^2, g^3, g^4 = 1$. Compare it with the multiplication table of the four-group. Are both groups abelian?

Exercise 3–41. Prove that a finite cyclic group is commutative.

Exercise 3–42. Prove that all cyclic groups of order e are isomorphic.

Exercise 3–43. Show how to represent a cyclic group of order e as an e-cycle, *i.e.*, a cyclic permutation on e letters.

Exercise 3–44. Write in cycle form the members of the cyclic group of order 8, and compare with the octic group. (In Exercise 3–19 you found the cycle forms for the octic group.) Are both groups abelian?

Quotient Groups

At this stage of our analysis we have a group G, a subgroup H of G, and the $[G:H]$ cosets of H in G, $g_1H, g_2H, g_3H, \ldots$. Can we do anything further with the cosets now that we have them?

In the abelian group of Figure 3–8 it appears that the cosets can be multiplied by multiplying their members with the resulting product another coset; in fact, if we multiply g^iH and g^jH in this way, the resulting coset is $g^{i+j}H$. To have the product of two cosets again be a coset for a nonabelian group G we shall need an extra condition on H. *What condition must we impose?*

In any group G with subgroup H, each member of the coset g_iH can be written g_ih for some member h of H. A member of the coset g_jH can be written g_jk for some $k \in H$. Multiplying them, we have

$$g_ihg_jk.$$

If this product is to lie in the coset g_ig_jH, then there must be a member m in H for which

$$g_ihg_jk = g_ig_jm.$$

G: cyclic group of order 12, $\{g, g^2, g^3, g^4, g^5, g^6, g^7, g^8, g^9, g^{10}, g^{11}, g^{12} = 1\}$, ·

H: cyclic group of order 2, $\{g^6, 1\}$, ·

Cosets: $\begin{Bmatrix} 1 \\ g^6 \end{Bmatrix}, \begin{Bmatrix} g \\ g^7 \end{Bmatrix}, \begin{Bmatrix} g^2 \\ g^8 \end{Bmatrix}, \begin{Bmatrix} g^3 \\ g^9 \end{Bmatrix}, \begin{Bmatrix} g^4 \\ g^{10} \end{Bmatrix}, \begin{Bmatrix} g^5 \\ g^{11} \end{Bmatrix}$

If we form a combination

$$\begin{Bmatrix} g \\ g^7 \end{Bmatrix} \cdot \begin{Bmatrix} g^3 \\ g^9 \end{Bmatrix} \rightarrow \begin{Bmatrix} g \cdot g^3 = g^4; & g^7 \cdot g^3 = g^{10} \\ g \cdot g^9 = g^{10}; & g^7 \cdot g^9 = g^{16} = g^4, \end{Bmatrix}$$

the result is $\begin{Bmatrix} g^4 \\ g^{10} \end{Bmatrix}$.

In fact, for cosets g^iH and g^jH we have

$$g^i\begin{Bmatrix} 1 \\ g^6 \end{Bmatrix} \cdot g^j\begin{Bmatrix} 1 \\ g^6 \end{Bmatrix} \rightarrow \begin{Bmatrix} g^i \\ g^{6+i} \end{Bmatrix} \cdot \begin{Bmatrix} g^j \\ g^{6+j} \end{Bmatrix} \rightarrow \begin{Bmatrix} g^ig^j = g^{i+j} & ; & g^{6+i}g^j = g^{6+i+j} \\ g^ig^{6+j} = g^{6+i+j}; & g^{6+i}g^{6+j} = g^{i+j} \end{Bmatrix}$$

$$= g^{i+j}\begin{Bmatrix} 1 \\ g^6 \end{Bmatrix},$$

a coset.

FIGURE 3–8

Multiply on the left by g_i^{-1} and on the right by k^{-1}:

$$g_i^{-1}(g_ihg_jk)k^{-1} = g_i^{-1}(g_ig_jm)k^{-1}$$

or, reassociating,

$$(g_i^{-1}g_i)(hg_j)(kk^{-1}) = (g_i^{-1}g_i)g_j(mk^{-1}),$$

so that

$$hg_j = g_jr,$$

where r is a member of H. Then each member of the right coset Hg_j is in the left coset g_jH.

Similarly, we can also show that the left coset g_jH is contained in the right coset Hg_j, so that the two must be equal: If multiplication of cosets is to be consistent, an element g_ih of g_iH times an element $g_j^{-1}s$ of $g_j^{-1}H$, $s \in H$, must lie in the coset $g_ig_j^{-1}H$. That is, there must be some $t \in H$ for which

$$g_ihg_j^{-1}s = g_ig_j^{-1}t.$$

Then

$$g_jg_i^{-1}(g_ihg_j^{-1}s)s^{-1}g_j = g_jg_i^{-1}(g_ig_j^{-1}t)s^{-1}g_j,$$

so that

$$g_jh = ts^{-1}g_j \in Hg_j.$$

This means that to have consistent multiplication of cosets of H in G we need to require that every left coset equal its corresponding right coset.

Definition 3–10.   A subgroup H of a group G is a **normal subgroup** of G if $gH = Hg$ for every element $g \in G$.

Notice that the notation $gH = Hg$ does not mean that g commutes with individual elements h of H. It means that for each gh_R there is some h_L in H for which $gh_R = h_Lg$. In an abelian group every subgroup is normal. In any group G the trivial subgroups G and $\langle 1 \rangle$ are normal.

Theorem 3–10.   The cosets of a normal subgroup H in G form a group with coset multiplication as operation.

PROOF: Coset multiplication is a binary operation on the cosets, for taking representatives of any two cosets g_iH and g_jH, we have, by associativity and the fact that $hg_j = g_jm$ for some $m \in H$, because H is a normal subgroup with $Hg_j = g_jH$,

$$(g_ih)(g_jk) = g_i(hg_j)k$$
$$= g_i(g_jm)k$$
$$= (g_ig_j)(mk).$$

We can verify associativity for coset multiplication, for

$$g_iH(g_jHg_kH) = g_iH(g_jg_kH) = g_i(g_jg_k)H,$$

and

$$(g_iHg_jH)g_kH = (g_ig_jH)g_kH = (g_ig_j)g_kH.$$

Thus associativity for coset multiplication follows from associativity of the group operation.

The identity coset is $1H = H$; the inverse coset of gH is $g^{-1}H$. ∎

Exercise 3–45. Let N be a normal subgroup of a group G. Prove that if n is an element of N, then $g^{-1}ng$ is an element of N for each element g of G.

Definition 3–11. Let G be a group with a normal subgroup H. The group of cosets of H in G is called the **quotient group or factor group G/H.** Alternatively, the elements of G/H may be thought of as the elements of G modulo H, so that two elements of the same congruence class, or coset, modulo H are equivalent.

The latter interpretation of G/H, as G taken modulo H, is essentially what we use in a finite cyclic group when we let the whole group $\langle g \rangle$ be represented by e distinct powers of G.

$$g, g^2, g^3, g^4, g^5, g^6 \equiv 1$$
$$g^4 \cdot g^5 = g^9 \equiv g^3$$

Example 1: *Clock Arithmetic.* On a 24-hour clock let the cyclic group $\langle g \rangle$ of order 24 represent the hours. Then each of the 24 numerals of the dial serves as representative of a whole coset of hours, a fact we use when we say that a 17 hour trip starting at 21 o'clock ends at $17 + 21 = 38 \equiv 14$ o'clock (the following day).

In shifting from a 24-hour clock to a 12-hour clock, we take the 24 hours modulo the normal subgroup $H: \{1, g^{12}\}, \cdot$. We can think of the resulting quotient group $\langle g \rangle/H$ as a group of 12 cosets or as a group of representatives, as:

$$g \sim \begin{cases} g \\ g^{13} \end{cases}; \quad g^2 \sim \begin{cases} g^2 \\ g^{14} \end{cases}; \quad g^3 \sim \begin{cases} g^3 \\ g^{15} \end{cases}; \dots; g^{11} \sim \begin{cases} g^{11} \\ g^{23} \end{cases}; \quad g^{12} \sim \begin{cases} g^{12} \\ g^{24} \end{cases}$$

An 8-hour trip beginning at 10 o'clock $\begin{cases} \text{a.m.} \\ \text{p.m.} \end{cases}$ ends at $\begin{cases} 6 \\ 18 \end{cases}$ or 6 o'clock $\begin{cases} \text{a.m.} \\ \text{p.m.} \end{cases}$. On the 12-hour clock we do not distinguish between the congruent elements in the same coset $\begin{cases} 6 \text{ a.m.} \\ 6 \text{ p.m.} \end{cases}$.

Exercise 3–46. Prove: In the (necessarily abelian) cyclic group $\langle g \rangle$ of order n, the normal subgroups are the cyclic groups $\langle g^d \rangle$, where d divides n evenly.

Exercise 3–47. Notice that among the integers I, the multiples of n form an infinite cyclic subgroup $\langle n \rangle$ under $+$. (Under the operation $+$, 0, n, $2n$, $3n$, and so forth play the roles of 1, g, g^2, g^3, and so forth under multiplication.) Since addition in I is commutative, any subgroup is normal. Consider $I/\langle n \rangle$. The cosets are

$$
\langle n \rangle = \left\{
\begin{matrix}
0 \\ +n \\ -n \\ +2n, \\ -2n \\ +3n \\ \cdot \\ \cdot \\ \cdot
\end{matrix}
\right.
\left\{
\begin{matrix}
1 \\ n+1 \\ -n+1 \\ 2n+1, \\ -2n+1 \\ 3n+1 \\ \cdot \\ \cdot \\ \cdot
\end{matrix}
\right.
\left\{
\begin{matrix}
2 \\ n+2 \\ -n+2 \\ 2n+2, \ldots, \\ -2n+2 \\ 3n+2 \\ \cdot \\ \cdot \\ \cdot
\end{matrix}
\right.
\left\{
\begin{matrix}
n-1 \\ 2n-1 \\ -1 \\ 3n-1. \\ -n-1 \\ 4n-1 \\ \cdot \\ \cdot \\ \cdot
\end{matrix}
\right.
$$

There are n distinct cosets. Notice, for instance, that the coset containing n is the same as the coset containing 0. We can use the n elements $\bar{0}, \bar{1}, \bar{2}, \ldots, n-1$ as representatives for the cosets in $I/\langle n \rangle$ with addition reduced modulo n. Complete the addition tables of $I/\langle 6 \rangle$ and $I/\langle 5 \rangle$ and show that $I/\langle n \rangle$ is a commutative group under addition modulo n.

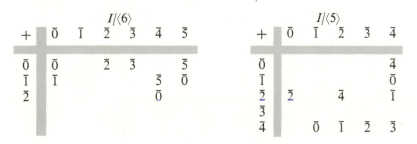

$I/\langle 6 \rangle$

+	$\bar{0}$	$\bar{1}$	$\bar{2}$	$\bar{3}$	$\bar{4}$	$\bar{5}$
$\bar{0}$	$\bar{0}$		$\bar{2}$	$\bar{3}$		$\bar{5}$
$\bar{1}$	$\bar{1}$				$\bar{5}$	$\bar{0}$
$\bar{2}$				$\bar{5}$	$\bar{0}$	

$I/\langle 5 \rangle$

+	$\bar{0}$	$\bar{1}$	$\bar{2}$	$\bar{3}$	$\bar{4}$
$\bar{0}$					$\bar{4}$
$\bar{1}$					$\bar{0}$
$\bar{2}$	$\bar{2}$		$\bar{4}$		$\bar{1}$
$\bar{3}$					
$\bar{4}$	$\bar{0}$	$\bar{1}$	$\bar{2}$	$\bar{3}$	

Example 2: *The Octic Group.*   In the octic group the subgroup V: $\{1, v\}$, $\circ$ is not normal, for $90 \circ v = d \notin V \circ 90$. The subgroup R of rotations is normal and G/R has cosets

$$\left\{ \begin{matrix} 1 \\ 90 \\ 180 \\ 270 \end{matrix} \right. , \quad \left\{ \begin{matrix} d \\ v \\ a \\ h \end{matrix} \right. ,$$

or representatives $1, d$ with multiplication taken modulo R. Physically, equating all symmetries that are congruent modulo R amounts to equating those mappings that differ only by rotations. We could do this by erasing the distinguishing numerals at the vertices and painting the front of the square, so as to equate all positions that involve just a rotation, while still detecting reflections (flips). The reflections are *involutory*, that is, of order 2. We can tell whether the square is in a reflected position from the original, but we cannot tell how many times it has been reflected, other than that it was reflected an odd number of times. If it appears unchanged it may have undergone any even number of reflections.

In Figure 3–6 we can see that the upper left and lower right quadrants of the multiplication table are made up of permutations of the members of R, while the upper right and lower left are made up of elements of the coset dR. We can write the multiplication table of G/R as:

coset multiplication	R	dR
R	R	dR
dR	dR	R

We can rearrange the table of Figure 3–6 so that the elements of the normal subgroup $H = \langle 180 \rangle$ appear first:

	1	180	90	270	d	a	v	h
1	1	180	90	270	d	a	v	h
180	180	1	270	90	a	d	h	v
90	90	270	180	1	h	v	d	a
270	270	90	1	180	v	h	a	d
d	d	a	v	h	1	180	90	270
a	a	d	h	v	180	1	270	90
v	v	h	a	d	270	90	1	180
h	h	v	d	a	90	270	180	1

Then the multiplication table of G/H is

	H	$90H$	dH	vH
H	H	$90H$	dH	vH
$90H$	$90H$	H	vH	dH
dH	dH	vH	H	$90H$
vH	vH	dH	$90H$	H

The formation of G/R in the octic group suggests the following theorem.

Theorem 3–11. Let H be a subgroup of a group G, with $[G:H] = 2$. Then H is normal and G/H is isomorphic to the cyclic group of order 2.

PROOF: If $h \in H$, then the coset $hH = H$ and the coset $Hh = H$, so $hH = Hh$. If $g \notin H$, then the coset gH has no member in common with H, for if s, say, were in both, we would have for some $t \in H$, $s = gt$, so that $g = st^{-1} \in H$. Similarly, the coset Hg has no members in common with H. H has $|H| = |G|/2$ members, and gH has the other $|G|/2$ members. Similarly, the coset Hg contains the other $|G|/2$ members not in H. Then gH and Hg are the same coset of H in G. Then, by Definition 3–10 H is normal in G.

Thus, the factor group G/H can be thought of as a group of cosets $\{H, gH\}$ under coset multiplication, where $g \notin H$. In Exercise 3–48 we justify one entry in the Cayley table

	H	gH
H	H	gH
gH	gH	H

This Cayley table is that of the cyclic group of order 2 with H as the identity element. ∎

Exercise 3–48. Show that in the preceding table $(gH)(gH)$ cannot equal gH, because $g \notin H$. Use this to show that $g^2 \in H$ for each $g \in G$.

Definition 3–12. Let S_n be the full symmetric group on n elements (Definition 3–2), with $n > 1$. Then the even permutations in S_n form a subgroup A_n, called the **alternating group on n elements** (see Exercise 3–26).

Corollary to Theorem 3–11. (This result is called a **corollary** rather than a theorem, because it is just a special case of Theorem 3–11.) The alternating group A_n is a normal subgroup of S_n for $n > 1$.

PROOF: To prove that $[S_n : A_n] = 2$ and so to apply Theorem 3–11 we need only show that all the odd permutations of S_n lie in the coset $(12)A_n$. Let an arbitrary odd permutation be expressed in transposition form as

$$(ij)a_n,$$

where a_n is an even permutation. Then

$$(ij)a_n = [(12)(12)][(ij)a_n] = (12)[(12)(ij)a_n],$$

and $(12)(ij)a_n \in A_n$. Then the index of A_n in S_n is 2, so by Theorem 3–11 A_n is a normal subgroup of S_n. ∎

It will be crucial to our study of solution by radicals that the symmetric group S_5 has as normal subgroup the alternating group A_5, but that A_5 has no proper normal subgroup.

Classroom Exercise 3–49. How many permutations are there in the symmetric group S_5? in the alternating group A_5?

Exercise 3–50. Complete the multiplication table of the quaternion group Q in Figure 3–9. Is the group abelian? Find its four proper subgroups and note that they are all normal subgroups.

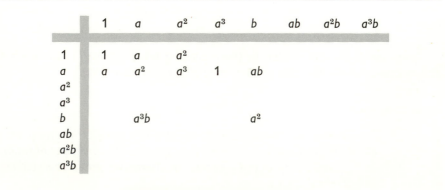

	1	a	a^2	a^3	b	ab	a^2b	a^3b
1	1	a	a^2					
a	a	a^2	a^3	1	ab			
a^2								
a^3								
b		a^3b				a^2		
ab								
a^2b								
a^3b								

FIGURE 3–9. The quaternion group Q.

Exercise 3–51. Write the multiplication table of the quotient group $Q/\langle a \rangle$, the quaternion group modulo its cyclic normal subgroup $\langle a \rangle$.

The reference in Appendix G shows another way to look at the quaternion group and other finite groups.

Exercise 3–52. Consider as a conjecture: "Let H be a subgroup of a finite group G. If H is abelian, then H is normal in G." Disprove this conjecture by reference to the octic group and its subgroup V.

Exercise 3–53. Let $K = \{1, v, h, 180\}$ be a subset of the octic group G. Show that $V = \{1, v\}$ is normal in K and that K is normal in G, but that v is not normal in G.

Exercise 3–54. Look up this generalization of Theorem 3–11: H. G. Bray, "A Generalization of a Result of Frobenius," American Mathematical Monthly, Vol. 76, No. 8, October 1969, pp. 924–925.

CHAPTER 4

MAPPINGS THAT
PRESERVE RELATIONS

Our immediate objective in this chapter is to develop a context in which the factor group G/H can be compared with the group G. The technique (of homomorphisms) to be used has much wider application in mathematics, so that the means becomes almost as important as the end.

To recapitulate, let G be a group, with H a normal subgroup, so that for each element $g \in G$, $gH = Hg$. (See Chapter 3 for the meaning of this symbolism.) Then there is a quotient group G/H, which can be thought of in two ways:

1. We can take the members of G/H to be the distinct cosets H, g_iH, g_jH, . . . and the group operation to be coset multiplication.

2. We can take the members of G/H to be members of G with congruent members identified, that is, with all congruent members of G considered equivalent, and the group operation to be multiplication modulo H.

How can we compare G/H with G? In (1) the elements of G/H are cosets, which are not elements of G. In (2) the operation of multiplication modulo H is not the same as the operation of G. So far we have available to us just one basis for comparing mathematical systems with different operations, namely isomorphism. Can we compare G and G/H on this basis? No, not in general, for isomorphism requires a one-to-one correspondence. If G is a finite group with $|G|$ elements and if H is a proper subgroup so that $|H| > 1$, then the order of G/H, which is $[G:H] = |G|/|H|$, is less than $|G|$. There is no way to establish a one-to-one correspondence between finite sets with unequal numbers of elements.

We are led by this to weaken the "one-to-one" part of the isomorphism Definition 2–15, retaining the "operation-preserving" feature. The result is called *homomorphism*.

Definition 4–1. Let S be a system of elements interrelated by a relation $Ⓡ$, and let T be another system with relation $\boxed{R}$. A **homomorphism** from S to T is a mapping ϕ of S to T such that elements of S that are $Ⓡ$-related map to elements of T that are $\boxed{R}$-related.

Figure 4–1 shows an art process, a rough version of the halftone reproduction technique used for newsphotos. The process illustrates a homomorphism from the set S of ink dots comprising the picture of Abel (from Figure HI–2) to the set T of crosses in the coarser silhouette. Transparent graph paper is laid over the dot picture. All the dots in any nonempty square of the graph paper are mapped to one cross covering that square. Let the relation $Ⓡ$ in S be "is at least as far down from the top of the picture as" and let the

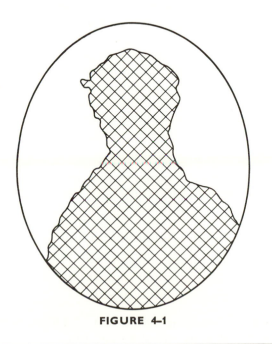

FIGURE 4–1

relation $\boxed{R}$ in T be "is at least as far down from the top of the silhouette as." Notice that the same mapping preserves various other relations of relative position.

A familiar mapping preserving certain relationships in the plane is the preparation of a cross-section or a flat projection of a 3-dimensional object. This kind of mapping can be considered a homomorphism when we are concerned only with those relations that it does preserve.

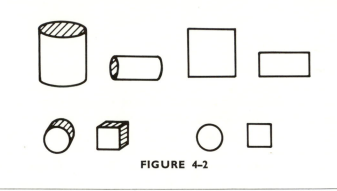

FIGURE 4–2

Generally, a homomorphism represents a lumping of information into grosser categories, and although individual data are sacrificed, items of data related by $\circledR$ contribute respectively to lumps of pooled data that are related by $\boxed{R}$.

As you realize, this is a very general and practical idea, which can be used to characterize any lumping of information, such as low-fidelity transmission of sound, black-and-white photography of colored objects, or rounding of numerical figures. Our need for homomorphisms in this chapter, though, centers on groups and their special relations, which are binary operations, so we include as Definition 4–2 a definition of homomorphism as a mapping from one group to another.

Classroom Exercise 4–1. Suppose that the numerical average grades G for a math class are lumped into letter grade categories L by a formula

$$90\text{--}100 \sim A$$
$$80\text{--}89 \sim B$$
$$70\text{--}79 \sim C$$
$$60\text{--}69 \sim D$$
$$\text{below } 60 \sim F$$

Is the correspondence a mapping from G into L? a homomorphism from G to L with respect to the relations "has at least as high a grade as" in G and "has a grade as early in the alphabet as" in L? with respect to "has a higher grade than" in G and "has a grade earlier in the alphabet than" in L? Replace the set L of five letters by the whole alphabet of 26 letters, and answer the questions again.

Classroom Exercise 4–2. Discuss whether the correspondence in Exercise 4–1 preserves the operation "averaging" as defined here for G and for L: In G, add the numerical scores, divide sum by the number of scores. In L, count 95 for each A, 85 for B, 75 for C, 65 for D, 30 for F, add, and divide by number of scores, convert to a letter grade by formula of Exercise 4–1.

> **Definition 4–2.** A **homomorphism from a group** G with binary operation $\times$ **to a group** $\bar{G}$ with binary operation $(\cdot)$ is a mapping ϕ that maps every $g \in G$ to some $\bar{g} \in \bar{G}$ so that the image of an $\times$-product in G is the corresponding $(\cdot)$-product in $\bar{G}$; that is,
>
> $$\phi(g_i \times g_j) = (\phi g_i) \cdot (\phi g_j).$$
>
> The set $\phi G \subseteq \bar{G}$ of all images $\bar{g} \in \bar{G}$ of elements $g \in G$ is called the **homomorphic image** of G. The various elements g sharing the common image $\phi g = \bar{g}$ make up the **pre-image** of $\bar{g}$. ϕ is a **homomorphism onto** $\bar{G}$ if the mapping is onto; that is, if $\phi G = \bar{G}$, so that every element of $\bar{G}$ appears as the image of at least one $g \in G$. If ϕ is onto and one-to-one, it is an isomorphism.

> **Theorem 4–1.** Let ϕ be a homomorphism from a group G to a group $\bar{G}$. Then the elements k of G for which $\phi k = \bar{1}$, the group identity of $\bar{G}$, form a normal subgroup K of G. (See Figure 4–3.)

PROOF: First we need to show that there are elements that belong to K; that is, that some elements do have the image $\bar{1}$. We can show that the identity 1 of G maps to $\bar{1}$: Since ϕ is a homomorphism we know by Definition 4–2 that

$$\phi 1 = \phi(1 \times 1) = \phi 1 \cdot \phi 1. \tag{1}$$

Since $\bar{G}$ is a group, the element $\phi 1$ has an inverse y in $\bar{G}$ for which $(\phi 1) \cdot y = y \cdot (\phi 1) = \bar{1}$, the $\bar{G}$-identity. Multiplying (1) by y gives us $\bar{1} = y \cdot \phi 1 = y \cdot (\phi 1 \cdot \phi 1) = (y \cdot \phi 1) \cdot \phi 1$ (by associativity) $= \bar{1} \cdot \phi 1$ (by the inverse property) $= \phi 1$ (by the identity property). Thus the image $\phi 1$ of the G-identity 1 is the $\bar{G}$-identity $\bar{1}$.

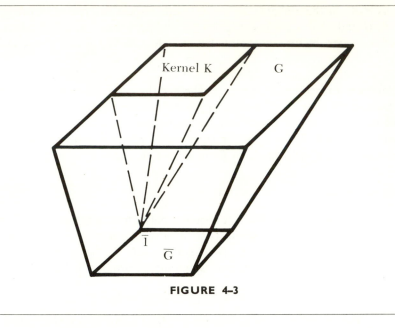

FIGURE 4–3

Now that we know K has members, we can show that it is a subgroup of G by showing that it has closure and the inverse property (see Theorem 3–6).

Closure: If k and m are in K, then

$$\phi k = \bar{1} \quad \text{and} \quad \phi m = \bar{1}.$$

Then $\phi(k \times m) = \phi k \cdot \phi m$ (by the homomorphism) $= \bar{1} \cdot \bar{1}$ (because $k, m \in K$) $= \bar{1}$ (by the identity property).

Inverse property: We have $\bar{1} = \phi 1 = \phi(k \times k^{-1})$ (by the inverse property) $= \phi k \cdot \phi(k^{-1})$ (by the homomorphism) $= \bar{1} \cdot \phi(k^{-1})$ (because $k \in K$), so that $\phi(k^{-1}) = \bar{1}$, or $k^{-1} \in K$.

This much proves that K is a subgroup of G. To prove that this subgroup is normal, we want to show that $gK = Kg$ for every g in G.

Let g be any element of G and let $\phi g = \bar{g}$ be its image in $\bar{G}$. Form the set M made up of all elements of G having this same image $\bar{g}$, that is, the pre-image of $\bar{g}$:

$$M = \{m \mid \phi m = \bar{g}\}.$$

All the products gk in gK are in M, for

$$\phi(gk) = \phi g \cdot \phi k = \phi g \cdot \bar{1} = \phi g = \bar{g}.$$

Similarly, all the products kg in Kg are in M, for

$$\phi(kg) = \phi k \cdot \phi g = \bar{1} \cdot \phi g = \phi g = \bar{g}.$$

Next, we prove that all the elements of M belong to gK. First, notice that the image of an inverse is the inverse of the image, for

$$\phi 1 = \phi(g \times g^{-1}) = \phi g \cdot \phi(g^{-1}) = \bar{g} \cdot \phi(g^{-1}) = \bar{1},$$

so that from the uniqueness of inverses (Theorem 1–2), we have

$$\phi(g^{-1}) = (\bar{g})^{-1}.$$

Then $\phi(g^{-1} \times m) = \phi(g^{-1}) \cdot \phi m = (\bar{g})^{-1} \cdot \bar{g} = \bar{1}$, so $g^{-1} \times m \in K$, and therefore $m \in gK$. Similarly, all the elements of M belong to Kg, for

$$\phi(m \times g^{-1}) = \phi m \cdot \phi(g^{-1}) = \bar{g} \cdot (\bar{g})^{-1} = \bar{1},$$

so that $m \times g^{-1} \in K$, and therefore $m \in Kg$.

We have proved that $M \subseteq gK \subseteq M$, which proves that $M = gK$, and that $M \subseteq Kg \subseteq M$, which proves that $M = Kg$. Together $M = gK$ and $M = Kg$ prove that $gK = Kg$. ∎

Definition 4–3. Let ϕ be a homomorphism from a group G to a group $\bar{G}$. The group K of pre-images of $\bar{1}$ is called the **kernel** of the homomorphism.

Theorem 4–1 and its proof remind us of the development of the quotient group G/H, in which the normal subgroup H is the identity. We tie the two together in a theorem.

Theorem 4–2. Let G be a group. The following two statements are equivalent; that is, each implies the other:

1. G has a normal subgroup K and factor group G/K.
2. There is a homomorphism from G onto a group $\bar{G}$ with kernel K.

PROOF: We already have all the essential parts of the proof. We note that if statement 1 applies to G, then we can define a mapping ϕ carrying each g into a coset gK. Then ϕ maps G onto $\bar{G} = G/K$. We proved in Theorem 3–10 that since K is normal, $(g_iK)(g_jK) = g_ig_jK$, which means that our mapping is a homomorphism. The kernel is K, the identity of G/K.

Now if statement 2 applies, we have from Theorem 4–1 that K is a normal subgroup of G, so by Theorem 3–10 there is a factor group G/K. ∎

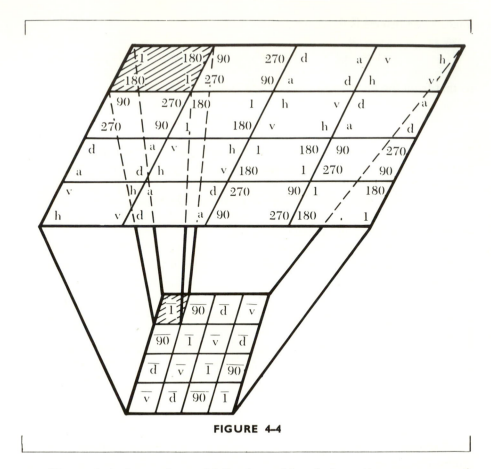

FIGURE 4–4

Figure 4–4 shows the multiplication table of the octic group mapped homomorphically onto $G/\langle 180 \rangle$, the octic group modulo its normal subgroup $\langle 180 \rangle$.

We introduced the octic group in Chapter 1 as the group of symmetries of a square, so of course the rotations and flips of a square form a physical model

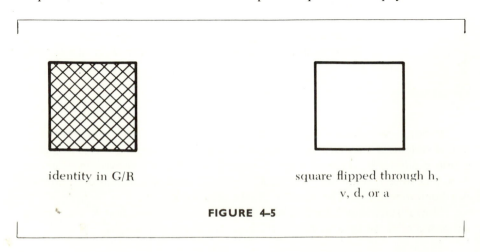

identity in G/R

square flipped through h, v, d, or a

FIGURE 4–5

for the octic group. Then in Chapter 3 we studied the factor group G/R of the octic group modulo its normal subgroup $\langle 90 \rangle$ of rotations. We were able to construct a physical model for the factor group by erasing the numerical labels of the vertices and painting the front of the square. In effect we "identified" the vertices in the sense that we made them all alike, no longer distinguishing between symmetries that differ only by a rotation.

Then we found the factor group $G/\langle 180 \rangle$ of the octic group modulo its normal subgroup $\langle 180 \rangle$ of rotations through multiples of 180°. The homomorphism from G onto $G/\langle 180 \rangle$ is illustrated in Figure 4–4. Is there a physical model for $G/\langle 180 \rangle$? Yes. Guided by the idea that in $G/\langle 180 \rangle$ rotations of multiples of 180° are to be alike, we identify vertices 1 and 4 and vertices 2 and 3 by shrinking the sides between them. Erase vertex labels and paint the front of the lozenge shape. Now individual vertices are not distinguished, but flips and rotations of 90° or 270° can be detected.

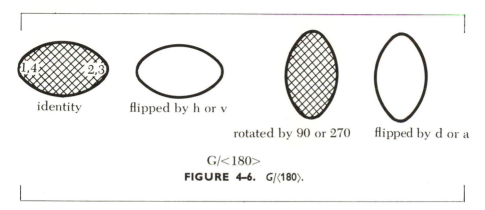

identity   flipped by h or v   rotated by 90 or 270   flipped by d or a

G/<180>

FIGURE 4–6. $G/\langle 180 \rangle$.

Now, in order that R serve as group identity in the quotient group G/R, we had to have $CR = RC$ for each coset C, and $CR = C$ for each coset C. By painting one side of the square and identifying all four vertices, we met the condition $CR = C$ by making rotations indistinguishable. Similarly, in $G/\langle 180 \rangle$, we can show that a coset is unaffected by a rotation of a multiple of 180° when we paint one side and identify pairs of adjacent vertices.

What about the condition that a group identity, even in a nonabelian group, must commute with each group element; does $CR = RC$? Does $D\langle 180 \rangle = \langle 180 \rangle D$ for cosets D of $\langle 180 \rangle$ in G? Yes, this condition is met because R and $\langle 180 \rangle$ are normal in G. How can this be interpreted in the physical models? In the model for G/R the condition is equivalent to the requirement that we cannot tell whether rotations have been made before or after a flip.

In the physical model for the whole octic group G we could not tell whether a position

arose from a horizontal flip h followed by a 90° rotation or from a 270° rotation followed by h. This is a physical realization of a *normal* subgroup, then. It is because such a condition was met that we were able to identify vertices to obtain a model for G/R. Similarly, the cosets of $G/\langle 180 \rangle$ in G were unaffected when preceded or followed by rotations in $\langle 180 \rangle$, and this made it possible to form a physical model for $G/\langle 180 \rangle$ by identifying adjacent pairs of vertices.

As you see, we have here a model for Theorem 4–2. The homomorphic image in G/R of

is

The homomorphic image in $G/\langle 180 \rangle$ of

1   2

□

4   3

is

We know such images must exist because the subgroups are normal (Theorem 4–2, part 2). Conversely, since the symmetries of

and

are groups, they have identity elements, which are homomorphic images of the respective kernels R and $\langle 180 \rangle$ in G, and the kernels must be normal subgroups of G (Theorem 4–2, part 1).

The symmetries of an icosahedron, a regular polyhedron with 20 equilateral triangles as faces, provide a physical realization for a group of permutations of five roots of a quintic (fifth degree) equation.

Exercise 4–3. Refer to the alternating group A_4 on four letters, whose multiplication table is shown in Figure 5–2. A homomorphism carries a, b^2, c, and d^2 to an element T of the cyclic group of order 3, and carries a^2, b, c^2, and d to T^2. Find the kernel K of the homomorphism.

Exercise 4–4. Show directly that K in Exercise 4–3 is a normal subgroup of A_4.

Exercise 4–5.   Find A_4/K in Exercise 4–3 and compare with $\langle T \rangle$.

Exercise 4–6.   Taking $\langle a \rangle$ as kernel, find a homomorphism from the four-group onto the cyclic group of order 2.

Exercise 4–7.   Taking $\langle a^2 \rangle$ as kernel, find a homomorphism from the cyclic group $\langle a \rangle$ of order 4 onto the cyclic group of order 2.

Exercise 4–8.   Refer to Exercise 3–47 about $I/\langle n \rangle$. The integers can be mapped onto the integers $0, 1, 2, \ldots, n-1$ modulo n, and that mapping is a homomorphism for the operation $+$. Is it also a homomorphism for $\times$? Compare ordinary multiplication as an operation on all the integers with multiplication modulo n among the integers modulo n. Let $\phi(i + rn) = i(\mathrm{mod}\ n)$, and show that $\phi[(i + rn)(j + sn)] = ij(\mathrm{mod}\ n)$.

Exercise 4–8 gives us an example of a homomorphism that preserves the two operations of a ring, mapping the ring of integers onto the ring of integers taken modulo n. This serves to introduce the following definition.

> **Definition 4–4.**   A homomorphism from a ring R into a ring S is a **ring homomorphism** if it preserves both ring operations, in R and S, respectively.

Classroom Exercise 4–9.   Let C be a finite cyclic group. Then prove that C is a homomorphic image of the integers I with operation $+$.

Classroom Exercise 4–10.   We have projected the octic group G onto its homomorphic images G/R and $G/\langle 180 \rangle$. Use Classroom Exercise 4–9 to suggest that the eight elements of G can themselves be considered homomorphic images of infinite classes of rotations and flips of the square.

Classroom Exercise 4–11.   Study Lemma L–1 in Appendix L. You will need to look up the definition of a solvable group.

CHAPTER 5

GROUPS OF PRIME ORDER; TWO ALTERNATING GROUPS

Among the positive integers N the *primes* are the multiplicative "atoms" or building blocks. They are the numbers 2, 3, 5, 7, 11, 13, and so forth as defined in Definition 5–1.

Definition 5–1. Among the positive integers the **primes** are the integers $p > 1$ for which

$$p \mid ab \quad \text{implies} \quad p \mid a \quad \text{or} \quad p \mid b,$$

which means "If p divides a product of integers ab, then either p divides a or p divides b."

The background algebra N is important, for if we allowed products of nonintegers, like $(\sqrt{3})(2\sqrt{3})$, then 3 could divide the product without dividing either factor. Notice that 1 is not a prime; as a divisor of 1, it is set aside in a class of its own as a **unit**. Positive integers that are not units and not primes are called **composite**. Primes are defined for other abstract algebras, such as all the integers I, by the same divisor property.

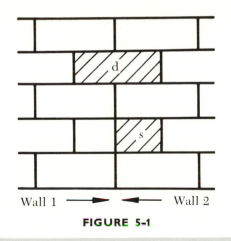

Wall 1 ➤   ◀ Wall 2

FIGURE 5-1

We might illustrate the way Definition 5–1 establishes primes as the basic building blocks of N. Figure 5–1 shows a set of building bricks, some double, some single square bricks. A double brick d (composite) may contribute to the building of two walls without contributing entirely to one of them. A single brick s (prime) must contribute entirely to one wall if it contributes to both combined.

We can show that primes are *irreducible*.

Definition 5–2. An integer $p > 1$ is **irreducible** if it has no proper divisors; that is, no divisors in N except 1 and p.

Theorem 5–1. In N the primes are irreducible.

PROOF: Let p be a prime in N and suppose that it is reducible, so that it factors as $p = mn$, with $1 < m < p$ and $1 < n < p$. Then $p \mid p$, but $p = mn$, so from Definition 5–1 $p \mid m$ or $p \mid n$, but $p > m$ and $p > n$, so this is impossible. We conclude that p is irreducible. ∎

It can also be proved that each irreducible integer is a prime, so that Definitions 5–1 and 5–2 are equivalent for N. (There are other systems in which some irreducibles are not primes; the equivalence applies to the integers I.) Either property could be used to define $2, 3, 5, 7, 11, \ldots$ among the integers, and the other property could then be deduced from it. It is the irreducibility property of the primes that we use in the following theorem about groups.

Theorem 5–2. Let G be a group of prime order p. Then G is cyclic.

PROOF: Since $p > 1$, G has some element $g \neq 1$, the identity of G. What is the order e of g, that is, the smallest positive integer for which $g^e = 1$ in G? From Theorem 3–9, $e \mid p$, the order of G. But p is irreducible by Theorem 5–1, so $e = 1$ or $e = p$. If $e = 1$, $g^1 = 1$, but $g \neq 1$, so $e = p$. Since $g, g^2, g^3, \ldots,$ and $g^p = 1$ all belong to G, these powers constitute the p elements of G, and G is cyclic. ▐

The following ingenious proof is due to James H. McKay, whose original report may be found in the American Mathematical Monthly, vol. 66 (1959), p. 119.

Theorem 5–3 (Cauchy). Let G be a finite group and let $p \mid |G|$, where p is a prime. Then G has a subgroup of order p.

PROOF: McKay proved that the number of elements in G that satisfy the equation $x^p = 1$ in G is a positive multiple kp of p. (Follow Figures 5–2 and 5–3 for a special case.) Then the deduction of the theorem is easy.

Form all possible lists of p elements chosen from G, repetitions allowed, whose product is 1:

$$(a_1 a_2 a_3 \cdots a_p),$$

where the product of the p elements listed is 1. The number of different lists is $|G|^{p-1}$, for we can choose any of the $|G|$ group members for $a_1, a_2, \ldots, a_{p-1}$, after which a_p is determined as $(a_1 a_2 \cdots a_{p-1})^{-1}$ since the product must be 1.

This is the multiplication table of A_4, the even permutations on 4 letters. Its order, $4!/2 = 12$, is divisible by 3, so by Theorem 5–3 A_4 should contain subgroups of order 3. By Theorem 5–2 these are cyclic subgroups. We.can see them by inspection of the multiplication table, but in Figure 5–3 we follow the details of the proof of Theorem 5–3 to illustrate it.

	1	a	a^2	b	b^2	c	c^2	d	d^2	r	s	t
1	1	a	a^2	b	b^2	c	c^2	d	d^2	r	s	t
a	a	a^2	1	t	c^2	d	s	r	b	c	b^2	d^2
a^2	a^2	1	a	d^2	s	r	b^2	c	t	d	c^2	b
b	b	s	c	b^2	1	t	d^2	a	r	c^2	d	a^2
b^2	b^2	d	t	1	b	a^2	r	s	c^2	d^2	a	c
c	c	b	s	r	d	c^2	1	t	a^2	a	d^2	b^2
c^2	c^2	r	d^2	a	t	1	c	b^2	s	b	a^2	d
d	d	t	b^2	c	r	s	a	d^2	1	a^2	b	c^2
d^2	d^2	c^2	r	s	a^2	b	t	1	d	b^2	c	a
r	r	d^2	c^2	d	c	b^2	a^2	b	a	1	t	s
s	s	c	b	a^2	d^2	a	d	c^2	b^2	t	1	r
t	t	b^2	d	c^2	a	d^2	b	a^2	c	s	r	1

Form lists like (111), (abt), (aaa), (bta), (rts), and so forth, each last element chosen to make the product of the three equal 1.

FIGURE 5–2

Now divide the lists $(a_1 a_2 \cdots a_p)$ into equivalence classes, calling two lists equivalent if one is a cyclic permutation of the other, i.e., $(a_1 a_2 \cdots a_p) \sim (a_{1+m} \cdots a_p a_1 \cdots a_m)$, $m \geq 0$. If in a list L_1 $a_1 = a_2 = \cdots = a_p$, then L_1 is the only list in its equivalence class. If in a list L_p there are at least two unequal members, $a_i \neq a_j$, then there are p lists in its equivalence class, i.e.,

$$(a_{1+m} \cdots a_p a_1 \cdots a_m)$$

for $m = 0, 1, \ldots, p - 1$. If there are u equivalence classes of one list each and v classes of p lists each, then the total number of lists is

$$u \cdot 1 + v \cdot p = |G|^{p-1}.$$

Now $p > 1$ and $p \mid |G|$, so $p \mid |G|^{p-1}$. Then since $p \mid vp$, we see that $p \mid u$, for $u = |G|^{p-1} - vp$. Since there is at least the L_1 list $(11 \cdots 1)$, we know u must be greater than zero. Then $u = kp$ for some positive integer k. That is, for kp elements a, $a^p = 1$. One such element is $a = 1$, but $p > 1$, so there must be an element $g \neq 1$ for which $g^p = 1$. Then, since p is irreducible, the cyclic subgroup $\langle g \rangle$ generated by g has order p. ∎

L_1	L_2

$$u \text{ classes} \atop \text{of 1 list} \atop \text{each} \left\{ \begin{matrix} (111) \\ (aaa) \\ (a^2a^2a^2) \\ (bbb) \\ \text{etc.} \end{matrix} \right. \qquad v \text{ classes} \atop \text{of 3 lists} \atop \text{each} \left\{ \begin{matrix} (abt), (bta), (tab) \\ (rts), (tsr), (srt) \\ (1tt), (tt1), (t1t) \\ (rst), (str), (trs) \\ \text{etc.} \end{matrix} \right.$$

total number of lists $= u \cdot 1 + v \cdot 3 = (12)^2$, so $u = 12^2 - 3v = 3(4 \cdot 12 - v)$. Then $3 \mid u$.

FIGURE 5-3

Exercise 5-1. For $p = 5$ find the five cyclic permutations of $(ababa)$. For $p = 4$ show that there are not four distinct cyclic permutations of $(abab)$.

Exercise 5-2. Which of these are primes in N? 57, 101, 322, 1011?

Exercise 5-3. Has A_4 any subgroup of order 4? (Refer to Figure 5-2.)

Exercise 5-4. Has A_4 any subgroup of order 6?

It can be proved that if p^r divides the order of a group, then the group has a subgroup of order p^r, where p is a prime.

Exercise 5-5. Show that S_5 has a subgroup of order 5. Find a permutation in S_5 that generates such a group.

We know from Theorem 4-2 that a group has no nontrivial quotient groups unless it has a proper normal subgroup. A group that has no proper normal subgroup is in a sense irreducible, like the prime integers, for it can have no nontrivial quotient group.

Definition 5-3. A group is **simple** if it has no proper normal subgroups.

The next result will contribute to our proof of Abel's Theorem that the general equation of degree greater than four is not solvable by radicals. Through Galois' structure theorem connecting fields and groups we can relate solution by radicals to the factoring of certain groups.

Theorem 5–4.   The alternating group A_5 is simple.

PROOF: A_5 stands for the group of even permutations on 5 letters. We shall suppose A_5 has a normal subgroup N, and prove that if $N \neq \langle 1 \rangle$, then $N = A_5$; that is, A_5 has only the two trivial normal subgroups $\langle 1 \rangle$ and A_5. This proof is *not* elegant. It is not especially hard, but since it involves checking several possible cases, it does not give us a feeling for structure, nor does it surprise us with some clever turn. However, it *does* prove the theorem, and the theorem can later take its place in a chain of reasoning.

First, notice that if n is one of the permutations in the normal subgroup N, then $a^{-1}na$ is also in N for every permutation $a \in A_5$ (see Exercise 3–45). Since N has the inverse property, it contains n^{-1}, so that by closure

$$a^{-1}nan^{-1} \in N \tag{1}$$

for every $a \in A_5$ and every $n \in N$. Consider those permutations $m \neq (1)$ in N that move the fewest letters, leaving the others fixed. Suppose each such m is written as a product of cycles with no letters in common (disjoint). Then all the cycles in the product have the same number of letters, for if

$$m = (\text{cycle of } r \text{ letters})(\text{cycle of } s \text{ letters}),$$

and $r < s$, consider m^r. The permutation m^r is in N by closure. It is not the identity permutation, because $s > r$, so the s letters moved by the second cycle have not returned to their original positions after r applications of the permutation m. But m^r moves fewer letters than m, for it does not move the r letters of the first cycle. Therefore, since m had a minimal property, we conclude that the disjoint cycles making up m have the same number of letters.

Exercise 5–6.   Express

$$\begin{pmatrix} 1 & 2 & 3 & 4 & 5 \\ 2 & 1 & 4 & 5 & 3 \end{pmatrix}$$

as a product of disjoint cycles.

Exercise 5–7.   If $m = (12)(345)$, find m^2 and show that it moves fewer "letters" than m.

Exercise 5–8.   If $m = (134)(2567)$. Find m^3 and show that it moves fewer letters than m.

i. Now suppose a minimal (moves fewest letters) permutation $m \in N$ has a cycle of 4 letters, say $m = (1234)$. (See Exercise 5–9 for more than 4 letters.)

We apply (1) with $a = (234)$ to show that N must also contain a permutation with a cycle

$$(234)^{-1}(1234)(234)(1234)^{-1}$$
$$= (432)(1234)(234)(4321)$$
$$= (124).$$

But (124) moves only 3 letters, which would contradict the minimal property of m. Using the same a, we can show that a 5-cycle (12345) is not minimal. Therefore m contains no single cycle of more than 3 letters. (See Exercise 5–9.)

Exercise 5–9. Suppose that in an alternating group A_x with $x > 5$ a minimal m contained a cycle (123456). Find $(234)^{-1}(123456)(234)(123456)^{-1}$, and show that it moves only 3 letters. Complete the proof for A_x of part *i*, by showing that no single cycle in m has more than 3 letters.

ii. In A_5 m cannot be a product of two or more disjoint 3-cycles, as only 5 letters are available.

Exercise 5–10. Suppose that in A_x for some $x > 5$ m has more than one k-cycle, $k \geq 3$. Suppose, for example, that in A_8 $m = (1234)(5678)$. Apply (1) with $a = (126)$ to give a 5-cycle in N, contrary to the minimal property assumed for m. Analogous constructions permit the conclusion that $m \in N$, a normal subgroup of A_x, $x > 5$, does not contain more than one k-cycle if m moves a minimal number of letters.

iii. Now suppose that a minimal m is made up of an even number of disjoint 2-cycles, *i.e.*, transpositions. The number must be even, because $m \in A_5$. Since the 2-cycles are disjoint, and we have only 5 letters available, there must be just two 2-cycles, such as $m = (12)(34)$. Apply (1) with $a = (125)$ to give

$$(521)[(12)(34)](125)[(34)(12)] = (125) \in N.$$

But (125) moves fewer letters than m, contrary to the minimal property assumed for m. Then a minimal m cannot be a product of 2-cycles.

Exercise 5–11. Prove *iii* for alternating groups of higher order than A_5 by applying (1) with $a = (123)$ to a product of an even number of 2-cycles beginning $m = (12)(34) \cdots$.

Now collecting the information from *i*, *ii*, and *iii*, we conclude that the minimal permutations in N (still excepting the identity permutation) are single 3-cycles, such as (rst). We assume that $N \neq \langle 1 \rangle$, so that N contains some permutations $\neq (1)$ and therefore at least one minimal permutation, of form (rst).

Next we show that since N contains a 3-cycle (rst), N must contain every 3-cycle. To show that it contains (123), for instance, we use the fact that N is normal to claim that N contains $a^{-1}(rst)a$, where a is one of the two permutations

$$\begin{pmatrix} 1 & 2 & 3 & 4 & 5 \\ r & s & t & u & v \end{pmatrix} \quad \text{or} \quad \begin{pmatrix} 1 & 2 & 3 & 4 & 5 \\ r & s & t & v & u \end{pmatrix},$$

whichever one is even and therefore a member of A_5. (Here r, s, t, u, v are the numerals 1, 2, 3, 4, 5 in some order.) Either computation gives $a^{-1}(rst)a = (123)$. We could adapt this argument to prove that any given 3-cycle is in N.

Further, we can show that every member of A_5 can be expressed as a product of 3-cycles, and thus by closure is a member of N: Each member of A_5 is an even permutation and therefore expressible as a product of an even number of transpositions, allowing this time for repetition of letters. Take the transpositions in adjacent pairs. If two transpositions are equal, they cancel each other, $(rs)(rs) = (1)$. If two transpositions have one letter in common their product is a 3-cycle, $(rs)(rt) = (rts)$. If two transpositions have no letter in common, their product can be expressed as the product of two 3-cycles, $(rs)(tu) = (tru)(rst)$. Then $A_5 \subseteq N \subseteq A_5$, so that if $N \neq \langle 1 \rangle$, then $N = A_5$. Therefore, A_5 has no proper normal subgroup N and so is simple. ∎

Exercise 5–12. Collect the results of Exercises 5–7 to 5–11 to show that the alternating group A_x for $x > 4$ is simple. ∎

Exercise 5–13. Find where the proof breaks down for $x = 4$.

Exercise 5–14. Refer to Exercises 3–47 and 4–8 about $I/\langle n \rangle$. Taking $0, 1, 2, \ldots, n - 1$ as representatives, write the multiplication tables of $I/\langle n \rangle$ for $n = 6$ and for $n = 5$.

Show that if n is composite, then $I/\langle n \rangle$ has proper divisors of zero and hence is not an integral domain (Definition 2–4) and hence also not a field.

Let p be a prime. Prove, using Definition 5–1, that the nonzero elements of $I/\langle p \rangle$ are closed under multiplication modulo p. Associativity is preserved in the homomorphism from the integers under multiplication to $I/\langle p \rangle$ under multiplication modulo p. The identity of $I/\langle p \rangle$ is 1 modulo p.

Prove that $I/\langle p \rangle$ has the inverse property by showing that for $1 \leq a < p$ the numbers, all $a \not\equiv 0 \pmod{p}$, $a \cdot 1, a \cdot 2, \ldots, a \cdot (p - 1)$ are distinct $\pmod{p}$ so that one of them, say $a \cdot y$ is congruent to 1 $\pmod{p}$. Then y is the inverse of a.

Prove the inverse property another way, by showing that if a is nonzero $\pmod{p}$, then so are its powers, which must repeat and are all in $I/\langle p \rangle$.

Collect results to show that $I/\langle p \rangle$ is a field (Definition 2–5).

Exercise 5–15. In Exercise 5–14 it was shown that the nonzero integers modulo p form a multiplicative group, of order $p - 1$. Prove that for any a not a multiple of p, $a^{p-1} \equiv 1$ modulo p. Prove that if q is a prime that divides $p - 1$, then the multiplicative group has a subgroup of order q. Show that the group of nonzero integers multiplied modulo 7 has a subgroup of order 3. In fact, find a, a^2, a^3, and a^6 for each $a \not\equiv 0 \pmod{7}$ and show that the number of these elements that satisfy $x^3 \equiv 1$ is a multiple of 3, and that the number that satisfy $x^2 \equiv 1$ is a multiple of 2.

HISTORICAL INTERMISSION

Much of the algebra covered in this book was developed by two young mathematicians early in the 19th century:

Niels Henrik Abel (1802–1829) of Norway

and

Evariste Galois (1811–1832) of France.

We can help picture the time in which they lived by recalling some dates:

1775 American Revolution

1789 French Revolution

1796 Napoleon took control of French Army

1812 Napoleon was defeated at Moscow

1814 Norwegian independence "in union with Sweden" was acknowledged

1830 Revolution in Paris

Some of the important mathematicians of the day:

France	Germany
1736–1813 Lagrange	1777–1855 Gauss
1749–1827 Laplace	1804–1851 Jacobi
1789–1857 Cauchy	1823–1891 Kronecker
1752–1833 Legendre	Ireland
1768–1830 Fourier	1805–1865 Hamilton
1788–1867 Poncelet	Russia
1822–1901 Hermite	1793–1856 Lobatchevsky

England

1815–1864 Boole

1814–1897 Sylvester

1821–1895 Cayley

Other famous creative people of the time:

1809–1882 Charles Darwin	1788–1824 Byron
1770–1827 Beethoven	1792–1822 Shelley
1797–1828 Schubert	1799–1850 Balzac
1809–1847 Mendelssohn	1749–1832 Goethe
1810–1856 Schumann	1775–1817 Jane Austen
1810–1849 Chopin	1812–1870 Charles Dickens
1811–1886 Liszt	1746–1828 Goya
1813–1883 Wagner	1819–1877 Courbet

The blanks are intended for your own write-in candidates. What was happening at this period in medicine? in architecture?

Evariste Galois

Galois (Gal-wah) was born just outside Paris, educated only by his mother until he was 12, then sent to Louis-le-Grand school in Paris. His sense of justice was continually outraged in this prison of a school (complete with bars at the windows), whose students and administrators reflected the political ferment of the times. He found nothing inspiring in the pedestrian artificial studies of rhetoric and classical literature that formed the backbone of the curriculum.

Finally it became Galois' turn to take the elementary mathematics course, which was to include a two-year study of Legendre's geometry. Young Galois read the whole geometry book like a novel and in that one quick reading mastered it.

He turned to algebra, but here ran into ordinary textbooks, not the shaped and polished complete elegance of Legendre's work. He discovered at about 14, as most creative mathematicians discover sooner or later, that he must go straight to originals. He ignored the required studies and gave himself instead a self-conducted reading course that would have taxed any of his teachers. He read his algebra directly from Lagrange, and he also read the papers of Abel.

Galois' development continued to diverge so widely from the academic course that there was never peace between him and the authorities whose approval he needed. He must have been a difficult pupil for an ordinary teacher—completely unresponsive to systems and requirements, on fire with mathematics to the exclusion of all else. It did not help that he did all of his work in his head, so that teachers conducting examinations or recitations were confronted only by conclusions. Since Galois worked only at the frontiers of mathematics as it was known in his day, those conclusions were often original and hard to understand.

If he was a difficult pupil, certainly he found his teachers difficult enough, too. Their arbitrary academic requirements made as little sense in Galois' individual case as it would to require a great poet to have flawless penmanship.

At 16 Galois failed the entrance examinations of the Ecole Polytechnique, for they concerned only the elementary and classical mathematics that Galois refused to study.

At 17 Galois did encounter a good teacher, in fact, an excellent one, the self-effacing Louis-Paul-Emile Richard, the advanced mathematics teacher at Louis-le-Grand. Richard gave full credit to Galois for his highly original solutions to difficult problems and encouraged his research, but he could not change his pupil's extraordinarily bad luck.

Cauchy had promised to present to the French Academy Galois' memoir containing the mathematics he had developed to date (that is, up to the age of 17). This included such new and comprehensive investigation of the theory of equations that a hundred years after Galois' death mathematicians were still filling in the details and seeing more applications. Cauchy forgot to submit the memoir and lost the abstract of it.

At 18 Galois again failed the entrance examinations for the Ecole Poly-technique, making examination history by hitting one of the examiners with a blackboard eraser, when he saw he had no chance to pass.

At 19 he submitted a memoir to the Academy of Science, with what seemed an excellent chance to take the Grand Prize in Mathematics. The Secretary took the manuscript home, but died before looking at it. It was never found.

Frustrated in every contact with academic life, Galois became active in political demonstrations. In May 1831 he was arrested for supposedly making a threat to the King's life during a banquet. A jury found him not guilty, but in July he was arrested as a "dangerous radical" and was later held prisoner six months on a trumped-up charge.

Apparently just after his release some of his political enemies rigged a duel with him, one that he foresaw he could not survive. All night before the duel he scribbled his mathematical will to the rest of us in a letter to his friend Chevalier—just an outline of his heroic accomplishments. Every now and then he wrote in the margin "I have not time. I have not time."

Early May 30, 1832 he was shot in the intestines and left to die. A peasant found him later and took him to the hospital, where Galois told his young brother, "Don't cry. I need all my courage to die at twenty." He died early the next morning and was buried in the common ditch of South Cemetery.

Niels Henrik Abel

Abel was one of seven children born to a poor pastor in the small village of Findö in Norway, a country so poor at the time that peasants emigrated in droves to escape starvation. Despite privation, the family remained happy, and Abel never seemed to resent his brothers and sisters, as he worked on mathematics with the equivalent of a TV set turned up full blast.

When Abel was about 15, his schoolteacher lost his post for beating a schoolboy fatally. The replacement teacher, Bernt Michael Holmboë (1795–1850), changed Abel's whole life. Abel learned the joys of a true mathematician, as he discovered for himself solutions to original problems Holmboë posed the class. By the age of 16 he was reading Newton, Euler, and Lagrange. He later explained that he advanced so fast "by studying the masters, not their pupils." Holmboë helped Abel master one of the most difficult and important memoirs in mathematics, the *Disquisitiones Arithmeticae* of Gauss.

When he was 18, the care of his mother and the six other children fell upon Abel, when his father died at 48. Abel hoped eventually to have a professorship, but eked out a living meanwhile with private pupils. He was recognized as a wonderful teacher himself. Holmboë, continuing to recognize his student's genius, befriended him, sometimes subsidized him from his own light pocket, worked constantly to get the Norwegian government to support Abel with grants so that he could travel and meet mathematicians from continental Europe. At 19 Abel completed the degree requirements at the University of Christiania (now Oslo; Denmark had permitted the establishment of a national university in 1811). The Norwegian government did scrape up enough for a small subsidy for Abel to continue studying at Christiania to improve his French and German. Finally when he was 23 he was granted enough to travel and study for one year in France and Germany.

Abel was deterred from trying to meet Gauss, because of the poor reception Gauss had given Abel's memoir on solution by radicals. He had tossed the paper aside unread, saying, "Here is another of those monstrosities!" In Berlin Abel met Crelle, who was just deciding whether to begin his mathematics journal, which continues to justify its fine reputation to this day. Abel supplied Crelle with excellent material, and Crelle supplied Abel with publication for his work.

Abel found the French mathematicians less approachable than the Germans. They received him civilly enough, but remained only polite. Abel took a trip south, leaving his comprehensive memoir with Cauchy for presentation to the Institut. Cauchy and Legendre were to have been the referees. Legendre was very old and complained in a letter to Jacobi that the manuscript was barely legible and written in pale ink. Cauchy took the manuscript home and lost it. E. T. Bell (*Men of Mathematics*, Dover, 1937) compared such forgetfulness to that of an Egyptologist mislaying the Rosetta Stone. Jacobi heard of the memoir's existence through Legendre, who had heard from Abel after his return to Norway. Jacobi wrote in 1829, "What a discovery is this of Mr. Abel's!... Did anyone ever see the like? But how comes it that this discovery that has been made in our Century, having been communicated to your Academy two years ago, has escaped the attention of your colleagues?" Finally diplomatic pressure from Norway led to a search, and Cauchy found the memoir in 1830, a year after Abel's death. Then someone lost the original manuscript while it was being printed so that the printed copy could not be proofread against the original.

Doctors in Paris told Abel that he had tuberculosis, but he refused to believe them. He returned to Berlin with the seven dollars he had left. Some of his grant money had not come through, so Holmboë lent him another sixty dollars, on which Abel lived and worked from March to May 1827 in Berlin, returning then to Norway. By early 1829 Abel realized that he was indeed dying of tuberculosis. Two days after his death in early April, Crelle wrote to say that he had at last wangled a professorship for Abel at the University of Berlin.

Some Comments on the Biographies

1. Was Abel the first to prove that the general fifth degree equation cannot be solved by radicals?

Abel thought for a while that he had found the algebraic solution for the general equation (a mistake Galois made later, too); then when he found his mistake, he decided such an algebraic solution was impossible and was able to prove it. To many the very idea of proving such a negative result was inconceivable; yet Paolo Ruffini (1765–1833) had offered several proofs from 1799, before Abel was born, to 1813. It is known that Abel saw some of Ruffini's work and commented that the reasoning did not always seem rigorous. By modern standards Ruffini's best proof glosses over an essential matter, but it does depend on substitution groups of the roots, as does Abel's. By modern standards Abel's proof needs a little retouching, too, though the essential chain of reasoning is there. Cauchy wrote to Ruffini in 1821 that Ruffini had "demontre completement l'insolubilite algebrique des equations generales d'un degre superieur au quatrieme." If he remained satisfied with Ruffini's proof, that may account for his negligence toward Abel's some five years later.

Rising standards of rigor in proofs make it possible for work even in mathematics to go out of date.

2. Which difficulties for Abel and Galois arose from communication lacks?

Again and again mathematicians have duplicated each other's work unnecessarily. Both Abel and Galois ran into difficulties with the referee systems of professional societies. A further problem, which caused them no worries because they died before it was released, was the diary of Gauss. Gauss did most of his mathematics for the joy of it, not for fame, and confided to his diary results that undercut the contributions of others for years to come, including some of Abel's and Jacobi's most important work on elliptic functions, Cauchy's on complex numbers, and Legendre's on the method of least squares.

A continuing problem in scientific communication is that difficulties in the subject matter produce difficulties in the exposition. Few can, like Galois, get the full import of an abstruse paper in one quick perusal. The newest most exciting original work is often the hardest to read, new to our thinking, flawed by little mistakes and gaps.

Is language itself a problem? English-speaking working mathematicians can usually read papers on their specialties in French and German. Increasingly there is need to know some Russian and Japanese. Abel had some trouble explaining his ideas orally in German and French when he first left Norway, but part of this was the novelty of those ideas. Nowadays the Scandinavian journals accept no papers in Scandinavian languages! In Gauss's time scientists could understand the concise Latin in which he wrote, provided they could understand the advanced mathematics he was describing.

You may find it surprisingly easy to get the gist of the opening paragraphs of a paper by Abel as it appeared in German and in French:

Beweis der Unmöglichkeit algebraische Gleichungen von höheren Graden als dem vierten allgemein aufzulösen.

(Von Herrn N. H. *Abel.*)

Bekanntlich kann man algebraische Gleichungen bis zum vierten Grade allgemein auflösen, Gleichungen von höhern Graden aber nur in einzelnen Fällen, und irre ich nicht, so ist die Frage:

Ist es möglich, Gleichungen von höhern als dem vierten Grade allgemein aufzulösen?

noch nicht befriedigend beantwortet worden. Der gegenwärtige Aufsatz hat diese Frage zum Gegenstande.

FIGURE HI–3

Démonstration de l'impossibilité de la résolution algébrique des équations générales qui passent le quatrième degré

On peut, comme on sait, résoudre les équations générales jusqu'au quatrième degré, mais les équations d'un degré plus élevé seulement dans des cas particuliers, et, si je ne me trompe, on n'a pas encore répondu d'une manière satisfaisante à la question:

"Est-il possible de résoudre en général les équations algébriques, qui "passent le quatrième degré?"

Ce mémoire a pour but de répondre à cette question.

FIGURE HI–3   (Continued)

3. Which work of Galois is represented in this book?

Galois treated the companion problem to Abel's, the problem of which equations are solvable by radicals. Abel showed that not all are solvable, but that does not mean all are not solvable. We are not going to study the companion problem, but we will rely on Galois' methods as we treat Abel's problem. Galois' methods are of the sort mathematicians call "beautiful" and "elegant." He invented such a natural way to describe the structure of the relations among the roots, that once we understand it, it seems to simplify and clarify all of theory of equations.

4. What can teachers learn from the lives of Galois and Abel?

One professor we know has never given a student a failing grade, lest he "flunk another Galois." What can a teacher do to discover and nurture mathematical talent? It helps to be prepared for it, to remind ourselves that real creativity is by its nature unconventional. We can never hope to progress if students are never smarter than their teachers or children smarter than their parents. For a hot-headed Galois we must provide recognition as early and often as possible, from commendations in class to prizes at science fairs. We must put him in communication with people who can understand him. Both Abel and Galois suffered at the hands of inept referees for their formal communications, but their personal correspondence with other mathematicians was full and satisfying. Abel was shy, Galois preoccupied. They probably didn't sell themselves very well to most audiences, but with pupils and colleagues they found communication easy.

Richard tried to help Galois and later was able to steer another difficult case, his famous student Hermite, through the Polytechnique examinations. (For all his brilliance, Hermite passed sixty-eighth from the top.) Abel's teacher Holmboë earned a hero's stature among teachers. He was adequate as a scholar himself but was not an innovator. Somehow, he could recognize what he had in Abel and, more mysterious still, continue to teach the boy more than either of them knew! How hard he must have worked outside school hours to understand and even correct Abel's early work! He read the masters right along with his pupil and was never left so far behind that he could not appreciate

Abel's output and provide an early sounding board for it. It is not uncommon for teachers who no longer sing to coach opera aspirants, but it is harder to imagine Holmboë, hanging onto the fast-moving Abel, steering that great power, but always from behind.

Few teachers encounter an Abel. We must only hope that the few who do will recognize the fact and bring such a student quickly into contact with working mathematicians. On the practical side, let teachers inform themselves of every resource, from school lunch programs and health clinics to scholarships, and let them not rely on formal rules when they have genius to deal with.

CHAPTER 6

POLYNOMIALS

Our information about polynomials seems at first quite heterogeneous, which makes us suspect that it is fragmentary. This is because we have learned things about polynomials in several different aspects and have not connected them in a coherent theory. Some of the types of information we want to connect here are

1. arithmetic of polynomials,
2. polynomial functions and graphing,
3. polynomial equations and roots.

The development of a theory of polynomials will provide us incidentally with a review of these various aspects.

Definition 6–1. Let R be a ring (Definition 2–1). The expressions

$$a_n x^n + a_{n-1} x^{n-1} + \cdots + a_1 x + a_0 = \sum_{i=0}^{n} a_i x^i,$$

where $n \geq 0$ is an integer and $a_i \in R$ for $i = 0, 1, \ldots, n$, form a ring $R[x]$, as we shall show in Theorem 6–1, called a **polynomial ring** under the operations

Addition:

$$\sum_{i=0}^{n} a_i x^i + \sum_{i=0}^{n} b_i x^i = \sum_{i=0}^{n} (a_i + b_i) x^i$$

Multiplication:

$$\sum_{i=0}^{m} a_i x^i \cdot \sum_{j=0}^{n} b_j x^j = \sum_{k=0}^{m+n} c_k x^k, \quad \text{where } c_k = \sum_{i+j=k} a_i b_j.$$

Each element of $R[x]$ is called a **polynomial**, and the R-elements a_i are called the **coefficients.** Two polynomials in $R[x]$ are equal if and only if their respective coefficients are equal in R. The **degree** **deg** P of a nonzero polynomial P is the maximum i for which $a_i \neq 0$. If the degree is d, then a_d is the **leading coefficient;** if R has a multiplicative identity 1 and $a_d = 1$, the polynomial is **monic.** The polynomials of zero degree, corresponding to the members of R, are called **scalars,** those of degree 1 **linear** polynomials, degree 2 **quadratic,** 3 **cubic,** 4 **quartic,** 5 **quintic.** The polynomial 0 is not assigned a degree.

Notice that we identify a polynomial not just by its appearance but also by its behavior, so that to define one polynomial we have to define a whole ring of polynomials with their operations $+$ and $\cdot$ and then define a single polynomial to be one element of this ring. In fact, a polynomial is not only a member of a certain set (of polynomial ring elements); it is a functioning element of the ring itself. This modern idea that an abstract algebra involves relations or operations as well as just members is fruitful and practical. As another example, recall that we do not define a separate integer, like 3, except as a functioning element among the natural numbers N, say, or all the integers I.

Exercise 6–1. Write out the polynomials $\sum_{i=0}^{4} a_i x^i$ and $\sum_{j=0}^{2} b_j x^j$.

Exercise 6–2. Express in summation notation

$$c_0 + c_1 x + c_2 x^2 + c_3 x^3 + c_4 x^4 + c_5 x^5.$$

Exercise 6–3. Find the sum and the product of the following polynomials over the ring of integers: $3 + x - 2x^2 + 4x^3$ and $23 + 5x - 7x^2$. To apply the addition rule, add a term $0x^3$ to the second polynomial, then use $n = 3$. Compare the use of the formula for multiplication with ordinary "long multiplication."

We must justify the assertion in Definition 6–1 that $R[x]$ is a ring.

Theorem 6–1. Let R be a ring. Then $R[x]$ is a ring. If R is a ring with (multiplicative) unit element, then $R[x]$ is a ring with unit element. If R is a commutative ring, then $R[x]$ is a commutative ring.

PROOF: For any two polynomials in $R[x]$ the sum and the product are in $R[x]$ according to the definitions of addition and multiplication, so $+$ and $\cdot$ are operations on $R[x]$. As in Exercise 6–3, to add polynomials of different degrees, we use the larger degree as n and introduce zero coefficients in the polynomial of smaller degree. The associative laws follow from the arithmetic of R, the ring of coefficients. The 0 of addition is the zero polynomial, which does not change any polynomial when added to it. The additive inverse of $\sum\limits_{i=0}^{n} a_i x^i$ is $\sum\limits_{i=0}^{n} (-a_i) x^i$. If R has a multiplicative identity 1, then the multiplicative identity in $R[x]$ is 1, the polynomial with $a_0 = 1$, $a_i = 0$ for $i > 0$, for it multiplies any polynomial in $R[x]$ without changing it. The left- and the right-distributive laws follow from the distributive laws in R.

Exercise 6–4. Verify the preceding sentence.

Exercise 6–5. Prove that if the multiplication in R is commutative, then the multiplication in $R[x]$ is commutative. ∎

If polynomials can be multiplied together, the product can be factored to give the original polynomials. This leads us to wonder which polynomials can be obtained as products of other polynomials from the same polynomial ring; that is, which polynomials can be factored? From experience with multiplication we learn to recognize some special products.

Exercise 6–6. Perform the multiplications

$(x + 2)^2$	$(x - 3)^2$	$5(x - 2)$
$-3(x + 4)$	$(x - 7)^2$	$(x + 4)(x - 4)$
$(x + 10)(x - 10)$	$(x + 1)(x - 1)$	$(x + 2)^3$
$(x - 1)(x^2 + x + 1)$	$(x + 1)^3$	$(x + 1)(x^2 - x + 1)$

Exercise 6–7. Factor the polynomials

$5x - 10$	$x^2 - 14x + 49$	$-3x - 12$
$x^2 + 4x + 4$	$x^2 - 6x + 9$	$x^3 + 3x^2 + 3x + 1$
$x^2 - 16$	$x^2 - 100$	$x^3 + 6x^2 + 12x + 8$
$x^2 - 1$	$x^3 - 1$	$x^3 + 1$

We have defined polynomials over a ring R with an arithmetic making them a ring. The power of x in each term of a polynomial serves only to mark the position of the coefficient in the polynomial, much as the power of 10 in decimal arithmetic dictates the position of each digit. A polynomial $9 - 2x + 3x^4$ could be written $(9, -2, 0, 0, 3)$, just as $3 + 9 \cdot 10 + 7 \cdot 10^4$ is written 70093.

The position marker x is called an *indeterminate*, and it does not take on values from R nor from any other set as does a variable. In the definitions of addition and multiplication the indeterminate continues to mark position, requiring us to add coefficients in like positions when adding polynomials and to indent and perform "long multiplication" when multiplying.

The indeterminate x in polynomials does not take on different values, but we can form a sum

$$s = \sum_{i=0}^{n} a_i r^i,$$

where r is a fixed element in the ring R, and the result, s, is a member of R, for R is closed under $+$ and $\cdot$. It would be desirable to have some kind of operation-preserving correspondence between polynomials $\sum_{i=0}^{n} a_i x^i$ and expressions like s, so that we could carry information from one system to the other. Does the natural correspondence $\sum_{i=0}^{n} a_i x^i \rightarrow \sum_{i=0}^{n} a_i r^i$ provide a homomorphism under $\cdot$? Consider the product of two polynomials $0 + 1x$ and $t + 0x$. From the multiplication rule for polynomials the product is

$$(0 + 1x)(t + 0x) = (0 \cdot t)x^0 + (0 \cdot 0 + 1 \cdot t)x^1 + (1 \cdot 0)x^2 = tx.$$

The mapping $\sum a_i x^i \rightarrow \sum a_i r^i$ maps the factors into $1r$ and t and maps the product into tr. If the mapping is to be a homomorphism with respect to $\cdot$ we must have

$$(1r)(t) = tr,$$

but in general in a ring R we cannot guarantee this, since r and t may not commute.

Although more general results can be established, and indeed are necessary for noncommutative rings, such as matrices, we now restrict our attention to polynomials over a field F, thus avoiding the problem of noncommutativity. Since a field is a ring, we can follow Definition 6–1 and construct a ring of polynomials in an indeterminate x with coefficients from the field F. In Theorem 6–1 we found that polynomials over a ring form a ring, polynomials over a ring with unit form a ring with unit, and polynomials over a commutative ring form a commutative ring. Do the polynomials over a field form a field? No, for nonzero polynomials of positive degree do not have inverse polynomials (see Classroom Exercise 6–8).

Classroom Exercise 6–8. Prove that the polynomial x in $F[x]$ has no multiplicative inverse in $F[x]$. (Multiply x by a polynomial $f(x)$ of degree

$d \geq 0$. What is the degree of the product? What is the degree of 1? Then deduce that $f(x)$ cannot be x^{-1}.)

Theorem 6–2. Let F be a field and $F[x]$ be the polynomial ring over F. Then $\phi: \sum_{i=0}^{n} a_i x^i \to \sum_{i=0}^{n} a_i d^i$, where d is a fixed field element, is a ring homomorphism from $F[x]$ into F.

PROOF: Does $\phi(P + Q) = \phi P + \phi Q$?

$$\phi(P + Q) = \phi\left(\sum_{i=0}^{n} a_i x^i + \sum_{i=0}^{n} b_i x^i\right) = \phi\left(\sum_{i=0}^{n} (a_i + b_i)x^i\right) = \sum_{i=0}^{n} (a_i + b_i)d^i,$$

while $\phi P + \phi Q = \sum_{i=0}^{n} a_i d^i + \sum_{i=0}^{n} b_i d^i$. We have to apply the associative and commutative laws of addition several times to write this as

$$\sum_{i=0}^{n} (a_i d^i + b_i d^i) = \sum_{i=0}^{n} (a_i + b_i)d^i,$$

by the distributive law. Then ϕ is a homomorphism under $+$.

Does $\phi(PQ)$ equal $\phi P \cdot \phi Q$? From the definition of multiplication

$$\phi(PQ) = \sum_{k=0}^{m+n} c_k d^k,$$

where $c_k = \sum_{i+j=k} a_i b_j$, while $\phi P \cdot \phi Q = \left(\sum_{i=0}^{m} a_i d^i\right) \cdot \left(\sum_{j=0}^{n} b_j d^j\right)$, which by several applications of the distributive law, the commutative laws for $+$ and for $\cdot$, and the associative laws for $+$ and for $\cdot$, can be reduced to $\sum_{k=0}^{m+n} c_k d^k$. Then ϕ is also a homomorphism under $\cdot$ and, hence, by Definition 4–4, a ring homomorphism. ∎

Exercise 6–9. Find the image of $3 + 7x - 4x^2$ under the homomorphism $\phi: \Sigma a_i x^i \to \Sigma a_i 5^i$. Find the image under the homomorphism $\psi: \Sigma a_i x^i \to \Sigma a_i (-5)^i$.

Exercise 6–10. Find the image of $2 - 3x + 5x^2 - 7x^3$ under a homomorphism that maps x to the complex number $2 + i$ ($i^2 = -1$).

> **Definition 6–2.** Let F be a field, and let a_i, $i = 0, 1, 2, \ldots, n$, be elements of F. A **polynomial function** $f(x)$ is a mapping f from F to F defined by
>
> $$f(d) = \sum_{i=0}^{n} a_i d^i.$$
>
> Any element mapped to 0 by f is called a **zero of $f(x)$.**

Note that $\sum_{i=0}^{n} a_i x^i$ is a polynomial in $F[x]$, so we can use Theorem 6–2 to apply everything we know about polynomial rings to the images $\sum_{i=0}^{n} a_i d^i$. Theorem 6–4 will give us a way to infer things about the polynomial ring from the images.

We write $f(x) = \sum_{i=0}^{n} a_i x^i$, notationally equating the polynomial function with its related polynomial, but when we use the function $f(x)$ it has the new feature that x is a *variable* that takes on values from F. Recall that in the polynomial ring $F[x]$ x was an indeterminate, merely holding position for the coefficients, and that the arithmetic of polynomials was not homomorphic to the arithmetic of functions of a variable x over a noncommutative ring.

Exercise 6–11. Given the polynomial function $3 + x^2$ over the field Q of rational numbers, find the images of $x = 0$, $x = 1$, $x = -1$, $x = a/b$, where $a/b \in Q$.

Exercise 6–12. Let $f(x) = -x^2 + 2x_4$ be a polynomial function over the real numbers. Is there a real number that is mapped into 0 by this function? Find the entire pre-image of 0.

Exercise 6–13. We established several facts about the arithmetic in $F[x]$ in Exercises 6–6 and 6–7. Reinterpret each of these facts as a fact about polynomial functions, and demonstrate the homomorphism for the cases $x = 0$, $x = 10$, and $x = \frac{1}{2}$. For instance, we found

$$x^3 - 1 = (x - 1)(x^2 + x + 1).$$

When $x = 10$ we expect from the homomorphism of Theorem 6–2 that the factoring will still hold:

$$10^3 - 1 = (10 - 1)(10^2 + 10 + 1).$$

If we cannot factor a polynomial by inspection, recognizing a familiar product, is there any other method of attack? One method is to divide the polynomial by a trial factor to determine whether it divides evenly.

Theorem 6–3 (Division Algorithm for $F[x]$). (An **algorithm** is a process that can be proved to solve a class of problems. This algorithm, for instance, will be proved to yield quotient and remainder when a polynomial in $F[x]$ is divided by another.) Let P and D be polynomials in $F[x]$, $D \neq 0$. Then there are a quotient polynomial Q and a unique remainder polynomial R such that

$$P = DQ + R,$$

and R is either 0 or of degree less than the degree of D.

PROOF: A formal proof of this theorem can be made by induction on the difference between the degree of P and the degree of D, but the notation is so elaborate as to obscure the basic idea instead of explaining it. Since the technique is the familiar one of long division, we start with an example and then call attention to the facts that make the algorithm work in general.

First, notice that if P has degree less than the degree of D, we can take Q to be the zero polynomial and $R = P$, satisfying $P = D \cdot 0 + P$, $\deg P < \deg D$. If $\deg P \geq \deg D$, we use long division, as in this example.

Let $P = 2x^4 - 5x^3 + x$, $D = 3x^3 - x + 4$.

$$
\begin{array}{r}
\frac{2}{3}x \;-\; \frac{5}{3} \\[4pt]
\hline
3x^3 - x + 4 \overline{)\,2x^4 \;-\; 5x^3 \qquad\qquad +\quad x} \\[4pt]
2x^4 \qquad\qquad -\; \frac{2}{3}x^2 \;+\; \frac{8}{3}x \\[4pt]
\hline
-\,5x^3 \;+\; \frac{2}{3}x^2 \;-\; \frac{5}{3}x \\[4pt]
-\,5x^3 \qquad\qquad +\; \frac{5}{3}x \;-\; \frac{20}{3} \\[4pt]
\hline
\frac{2}{3}x^2 \;-\; \frac{10}{3}x \;+\; \frac{20}{3}
\end{array}
$$

Then $Q = \frac{2}{3}x - \frac{5}{3}$ and $R = \frac{2}{3}x^2 - \frac{10}{3}x + \frac{20}{3}$, and we have $2x^4 - 5x^3 + x = (3x^3 - x + 4)(\frac{2}{3}x - \frac{5}{3}) + (\frac{2}{3}x^2 - \frac{10}{3}x + \frac{20}{3})$, with $\deg R = 2 < 3$.

At each stage the division is possible, for the leading coefficient, say b, in D is a nonzero field element and so has an inverse b^{-1}. If the leading coefficient in the remainder so far is a field element c, then the new coefficient in the quotient is $b^{-1}c$.

The division goes on until the degree of the latest remainder is less than the degree of D, for if at some stage the remainder so far has too great a degree, we divide again.

To show that R is unique, suppose $P = DQ_1 + R_1$ and also $P = DQ_2 + R_2$. Then $DQ_1 + R_1 = DQ_2 + R_2$, which implies that $D(Q_1 - Q_2) = R_2 - R_1$, so that D divides $R_2 - R_1$. But $\deg R_i < \deg D$ or 0, for $i = 1, 2$, so $R_2 - R_1 = 0$ and the remainders are equal. ∎

Exercise 6–14. Divide $x^5 + 4x^4 + 1$ by $x^3 + 7$ over the field Q of rational numbers.

Exercise 6–15. Factor $x^5 - x^3 - x^2 - 6x - 2$ with $x^2 + 2$ as one of the factors.

Exercise 6–16. Determine which of the following linear polynomials are factors of the polynomial $x^5 - x^4 - 2x^3 + 2x^2 - 15x + 15$ over the real field.

$$x - 1 \qquad x - \sqrt{5} \qquad x + \sqrt{5}$$

$$x + 1 \qquad x - \sqrt{3} \qquad x + \sqrt{3}$$

The following theorem provides us with a quick way to check for linear factors of a polynomial by finding which field elements are mapped to 0 by the polynomial function.

Theorem 6–4 (Factor Theorem).   Let P be a polynomial in $F[x]$, where F is a field.   Then $x - d$, for $d \in F$, is a factor of $P = f(x)$ if and only if d is a zero of the polynomial function $f(x)$.

PROOF: Apply Theorem 6–3 with $D = x - d$. Then the remainder R is either 0 or of degree $<$ deg $D = 1$. Then the remainder is a field element r.

$$P = f(x) = [Q(x)](x - d) + r,$$

where $r \in F$. Then the homomorphism in Theorem 6–2 insures that

$$f(d) = [Q(d)](d - d) + r.$$

Then if d is mapped to 0 by $f(x)$, so that $f(d) = 0$, we have

$$0 = f(d) = [Q(d)](0) + r,$$

so that $r = 0$.
  Conversely, if $x - d$ is a factor of P, so that $r = 0$, then

$$f(d) = [Q(d)](0) = 0,$$

so that $f(x)$ maps d to 0.   ∎

Exercise 6–17. Referring to Exercise 6–16 verify that in each case $x - d$ is a factor of the given polynomial if and only if

$$d^5 - d^4 - 2d^3 + 2d^2 - 15d + 15 = 0.$$

Exercise 6–18. Use Theorem 6–4 to determine which of the following linear polynomials are factors of $2x^4 + 11x^3 + x^2 + 121x - 231$: $x - 1$, $x + 7$, x, $x - \sqrt{11}$, $2x - 3$ (note that $2x - 3 = 2(x - \frac{3}{2})$).

Exercise 6–19. Complete the statement and proof for a corollary to Theorem 6–4 covering the special case $d = 0$: "Corollary. x is a factor of $P = \sum\limits_{i=0}^{n} a_i x^i$ if and only if $a_0 = \cdots$."

Exercise 6–20. Refer to Exercise 6–19 and generalize the corollary to cover the case x^k, where $k > 0$. How large can k be?

Definition 6–3. A **polynomial equation** is an equation

$$f(x) = 0,$$

where $f(x)$ is a polynomial function. The **roots** of the equation are the zeros of the function, that is, the values that are mapped to 0 by $f(x)$.

We often take advantage of the fact that the roots of $f(x) = 0$ are the zeros of $f(x)$. Suppose we want to solve the equation

$$2x^3 + 3x^2 - 29x + 30 = 0;$$

that is, we want to find all the roots r for which

$$2r^3 + 3r^2 - 29r + 30 = 0.$$

We might try a few values to see whether they are roots, starting with values that make the calculations easy, like 0, 1, and -1.

$$2(0)^3 + 3(0)^2 - 29(0) + 30 = 30 \neq 0$$
$$2(1)^3 + 3(1)^2 - 29(1) + 30 = 6 \neq 0$$
$$2(-1)^3 + 3(-1)^2 - 29(-1) + 30 = 60 \neq 0$$

Although these first three trials have not yielded roots, we need not waste the calculations. We can graph the function $f(x)$ on Cartesian coordinate paper as in Figure 6–1 and use its nonzero values to indicate where zeros can be expected to lie.

The next theorem enables us to shorten the work of calculating $f(d)$ for a trial value d. It is a generalization of the Factor Theorem 6–4.

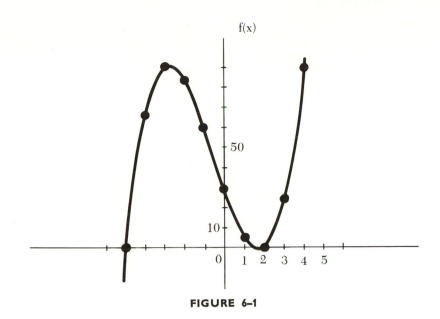

FIGURE 6–1

Theorem 6–5 (Remainder Theorem). Let $P = f(x)$ be a polynomial (function) with coefficients from a field F. Then the function value $f(d)$ for a field element d equals the remainder when the related polynomial P is divided by $x - d$.

PROOF: From the Division Algorithm, Theorem 6–3, there exist a quotient polynomial Q and a remainder polynomial R with $R = 0$ or $\deg R < 1$, for which

$$P = [Q(x)](x - d) + R.$$

Since $R = 0$ or $\deg R < \deg(x - d) = 1$, R is a scalar, or constant, r, an element of the field F. Now apply the homomorphism ϕ of Theorem 6–2.

$$f(d) = [Q(d)](d - d) + r,$$

or

$$f(d) = r. \quad \blacksquare$$

Exercise 6–21. Show that Theorem 6–4 is a corollary to Theorem 6–5.

In Figure 6–2 we use Theorem 6–5 to find $f(-2)$ when

$$(x) = 3x^3 - 25x + 1.$$

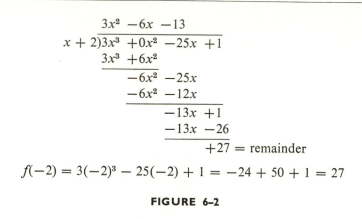

$$f(-2) = 3(-2)^3 - 25(-2) + 1 = -24 + 50 + 1 = 27$$

FIGURE 6–2

As you may recall, synthetic division is a shorthand for the division process obtained by leaving out the position-holding powers of the indeterminate x, relying on position alone. The division in Figure 6–2 would appear in synthetic division as in Figure 6–3.

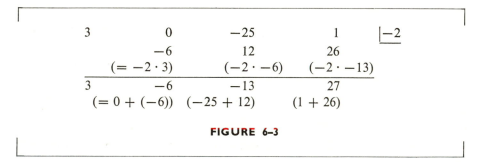

FIGURE 6–3

The numbers in the last row in Figure 6–3 yielded by the process are reinterpreted according to their positions as coefficients of the quotient polynomial and the remainder:

$$3 \quad -6 \quad -13 \quad 27 \rightarrow 3x^2 \quad -6x \quad -13, \text{ remainder } 27$$

Figure 6–4 illustrates the use of the remainder theorem to find function values and locate zeros.

The graph in Figure 6–4 suggests that as x increases beyond 3 the function continues to increase beyond 138, so that there are no zeros of the function for $x > 3$. Analysis of the synthetic division confirms this, as all quotient coefficients and remainder are positive (last line of the division). Study of the synthetic division for negative x confirms that the function remains negative for $x < -1$; entries in the last row of the division alternate in sign, and for $x < -1$ the remainder is negative. The second derivative has just one real zero, at the positive cube root of $\frac{3}{2}$, where the first derivative stops decreasing and starts increasing with larger values of x. We can deduce from this kind of argument (or directly from "Descartes' Rule of Signs"; see J. V. Uspensky,

$f(x) = x^5 - 15x^2 + 9x + 3$
$f'(x) = 5x^4 - 30x + 9$
$f''(x) = 20x^3 - 30$

x	f(x)	f'(x)
0	3	9
1	-2	-16
-1	-22	44
2	-7	29
3	138	
0.5	3.78	-5.6

```
1    0   0   -15    9     3   |2
         2   4     8   -14  -10
     _____
1    2   4    -7   -5    -7

1    0   0   -15    9     3   |3
         3   9     27   36   135
     _____
1    3   9    12   45   138

1    0   0   -15    9     3   |-1
        -1   1     -1   16   -25
     _____
1   -1   1   -16   25   -22
```

$$f(x) = x^5 - 15x^2 + 9x + 3$$

Theory of Equations, McGraw-Hill, 1948, pp. 121–124) that the equation
$x^5 - 15x^2 + 9x + 3 = 0$ has exactly three real roots, one negative and two
positive.

Exercise 6–22. Analyze and graph the function $f(x) = x^5 - 4x^4 + 2x + 2$,
following the example in Figure 6–4. Find the first and second derivatives, and

$f(x) = x^5 - 15x^2 + 9x + 3$

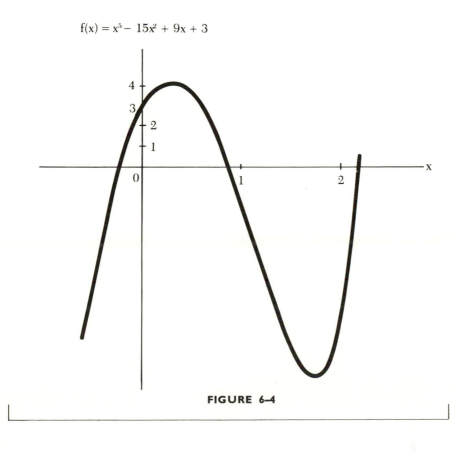

FIGURE 6–4

recall from calculus (Rolle's Theorem) that $f''(x)$ has a zero between successive zeros of $f'(x)$. Deduce that the equation $x^5 - 4x^4 + 2x + 2 = 0$ has exactly three real roots. In Theorem 10–1 we shall prove that the roots of this equation cannot be found "by radicals."

Factoring Polynomials in $I[x]$

Theorem 6–6 (Lemma of Gauss). (A **lemma,** or etymologically a "horn" [cf. "dilemma," two-horn], of a proof is a result established not so much for its own sake as for its contribution to proving a theorem.) Let $f(x) = \sum\limits_{i=0}^{n} a_i x^i$ be a polynomial in $I[x]$. Let the only common divisors of $a_0, a_1, \ldots,$ and a_n be 1 and -1. If $f(x)$ can be factored in $Q[x]$

$$f(x) = g(x) \cdot h(x),$$

where the coefficients of g and h are rational numbers, then $f(x)$ can be factored in $I[x]$.

PROOF: Let the rational coefficients in the factors $g(x)$ and $h(x)$ be written in lowest terms. (Follow Figure 6–5.) Let the least common multiple of the denominators of the g-coefficients be u and express $g(x)$ as $(1/u)\bar{g}(x)$, where $\bar{g}(x)$ is a polynomial with integral coefficients. Then factor out the greatest common divisor s of the coefficients of $\bar{g}(x)$, writing $g(x)$ as $(s/u)G(x)$, where the (integral) coefficients of $G(x)$ have only ± 1 as common factors. Let v and t play the roles of u and s, respectively, for the polynomial $h(x)$, so that $h(x) = (1/v)\bar{h}(x) = (t/v)H(x)$, where $H(x)$ has integral coefficients with no common factors except ± 1. Then

$$f(x) = \frac{s}{u} G(x) \cdot \frac{t}{v} H(x) = \frac{st}{uv} G(x)H(x),$$

where $G(x)$ and $H(x)$ are in $I[x]$ and s, t, u, and v are integers.

$$6x^4 - x^3 - 8x^2 + 32x - 120 = (2x^2 - \tfrac{4}{3}x + 8)(3x^2 + \tfrac{3}{2}x - 15)$$
$$= \frac{6x^2 - 4x + 24}{3} \cdot \frac{6x^2 + 3x - 30}{2}$$
$$= \tfrac{2}{3}(3x^2 - 2x + 12) \cdot \tfrac{3}{2}(2x^2 + x - 10)$$
$$= \tfrac{2}{3} \cdot \tfrac{3}{2}(3x^2 - 2x + 12)(2x^2 + x - 10)$$

FIGURE 6–5

But we can show that $st = \pm\, uv$: In the equation

$$uvf(x) = stG(x)H(x)$$

the only common factor of the coefficients of $uvf(x)$ is uv. Hence $st \mid uv$, since st divides the whole left member. We can also show that no prime p can divide all the coefficients of the product polynomial $G(x)H(x)$; therefore we will be able to prove that $uv \mid st$. From our construction, any prime p fails to divide some coefficient of $G(x) = \sum_{j=0}^{b} g_j x^j$. Let g_y be the first coefficient that p fails to divide. Then p divides $g_0, g_1, \ldots$, and g_{y-1}, but $p \nmid g_y$. Similarly, $H(x) = \sum_{k=0}^{c} h_k x^k$. Let p divide $h_0, h_1, \ldots$, and h_{z-1}, but $p \nmid h_z$. The coefficient of x^{y+z} in the product $G(x)H(x)$ is

$$(g_{y+z}h_0 + \cdots + g_{y+1}h_{z-1}) + g_y h_z + (g_{y-1}h_{z+1} + \cdots + g_0 h_{y+z}),$$

where all the terms in the left parentheses have factors h_k with $k < z$ and so are divisible by p, and all the terms in the right parentheses have factors g_j with $j < y$ and so are divisible by p. However, $p \nmid g_y h_z$, so $p \nmid G(x)H(x)$. Then since the only common factors of all coefficients in $stG(x)H(x)$ are $\pm st$, we see that $uv \mid st$. Then

$$f(x) = \pm G(x)H(x). \quad \blacksquare$$

Exercise 6–23.   Factor

$$15x^5 - 9x^4 - 4x^3 + 21x^2 - 13x + 6 = (5x^3 - \tfrac{10}{3}x + \tfrac{15}{3})(3x^2 - \tfrac{9}{5}x + \tfrac{6}{5})$$

in $I[x]$.

Definition 6–4. A polynomial $f(x)$ of positive degree is **irreducible** in $F[x]$ if it has no factorization

$$f(x) = g(x)h(x)$$

with $g(x)$ and $h(x)$ in $F[x]$ of positive degrees. If $f(x) = g(x)h(x) \in F[x]$, then we say $g(x)$ divides $f(x)$, or in symbols $g(x) \mid f(x)$, and $h(x) \mid f(x)$.

We must emphasize that the irreducibility of a polynomial depends on the field. For instance, we have seen that $x^2 + 1$ is irreducible over $I[x]$, $Q[x]$, and $R[x]$ but factors into $(x + i)(x - i)$ in $C[x]$. In fact, we custom-made the field C of complex numbers to make the factorization possible.

Exercise 6–24. The integers taken modulo a prime p form a field $I/\langle p \rangle$ (cf. Exercise 5–14). Show that in $(I/\langle 2 \rangle)[x]$ the polynomial $x^2 + 1$ factors, because $x^2 + 1$ has a zero at $x = 1$. Show that in $(I/\langle 3 \rangle)[x]$ the polynomial $x^2 + x + 1$ factors. Show that in $(I/\langle p \rangle)[x]$ the polynomial $x^{p-1} + \cdots + x^2 + x + 1$ factors.

Theorem 6–7 (Eisenstein's Irreducibility Criterion). Let $f(x) = \sum_{i=0}^{n} a_i x^i$ be a polynomial in $I[x]$, whose coefficients a_i for $i = 0, 1, 2, \ldots, n-1$ are divisible by a prime p but $p \nmid a_n$ and $p^2 \nmid a_0$. Then $f(x)$ is irreducible in $Q[x]$.

PROOF: From Theorem 6–6 we need only prove that $f(x)$ does not factor in $I[x]$, for if it factored in $Q[x]$ it would factor in $I[x]$. Suppose $f(x) = g(x)h(x)$ and $\deg g(x) > 0$, $\deg h(x) > 0$. Write $g(x) = \sum_{i=0}^{k} g_i x^i$, $h(x) = \sum_{i=0}^{m} h_i x^i$, where the coefficients g_i and h_i are integers (follow Figure 6–6). The prime p does not divide $g_k h_m$, for $g_k h_m = a_n$, so $p \nmid g_k$ and $p \nmid h_m$. Also, from hypothesis, we have $p \mid a_0$ but $p^2 \nmid a_0$; however, $a_0 = g_0 h_0$. Then from Definition 5–1 of primes we know $p \mid g_0$ or $p \mid h_0$, and from $p^2 \nmid a_0$ only one of g_0 and h_0, say g_0, is divisible by p.

Suppose $p \mid g_0, p \mid g_1, \ldots, p \mid g_{s-1}$, but $p \nmid g_s$. Compute a_s, the coefficient of x^s in the product polynomial. Since $0 < s \le k < n$, $p \mid a_s$, where

$$a_s = g_s h_0 + g_{s-1} h_1 + \cdots + g_1 h_{s-1} + g_0 h_s.$$

Take $p = 5$.

$$5 - 25x^3 + 2x^5 = (g_0 + g_1 x + g_2 x^2 + g_3 x^3)(h_0 + h_1 x + h_2 x^2)$$

$$2 = g_3 h_2,$$

so $5 \nmid g_3$ and $5 \nmid h_2$.

$$5 = g_0 h_0,$$

so $5 \mid g_0$ or $5 \mid h_0$, but not both. Suppose $g_0 = 5\bar{g}_0$, $g_1 = 5\bar{g}_1$, but $5 \nmid g_2$. Compute $a_2 = g_2 h_0 + g_1 h_1 + g_0 h_2$, obtaining

$$0 = g_2 h_0 + 5\bar{g}_1 h_1 + 5\bar{g}_0 h_2,$$

from which $5 \mid g_2 h_0$, but $5 \nmid g_2$ and $5 \nmid h_0$, contradicting Definition 5–1, since 5 is a prime.

FIGURE 6–6

Then p divides $g_s h_0 = a_s - (g_{s-1} h_1 + \cdots + g_0 h_s)$. But $p \nmid g_s$ and $p \nmid h_0$, contrary to Definition 5-1. We have reached a contradiction. There is no choice of coefficients for factors $g(x)$ and $h(x)$ that will yield the divisibility properties assumed in the hypothesis. ∎

Exercise 6–25. Show that $x^3 + 26x^2 - 52x + 13$ is irreducible in $Q[x]$.

Exercise 6–26. Show that $x^2 - c$ is irreducible over $Q[x]$ if $c > 1$ is not divisible by a square > 1. Show that if c is not a perfect square then $x^2 - c$ is irreducible.

Exercise 6–27. Write two fifth degree polynomials that can be shown by Theorem 6–7 not to factor in $Q[x]$.

Exercise 6–28. Prove that the polynomial of Exercise 6–22 is irreducible in $Q[x], f(x) = x^5 - 4x^4 + 2x + 2$.

Corollary. Let p be a prime. The polynomial $x^{p-1} + \cdots + x^2 + x + 1$ is irreducible in $Q[x]$.

PROOF: In Exercise 6–24 we showed that this polynomial factors in $I/\langle p \rangle [x]$, because $1 + \cdots + 1 + 1 + 1 = p \equiv 0$ modulo p, so that

$$(x - 1) \mid (x^{p-1} + \cdots + x^2 + x + 1)$$

modulo p.

However, we can recast the polynomial so as to bring it within the scope of the Eisenstein Irreducibility Criterion (Theorem 6–7) and so prove that it does not factor in $Q[x]$. Writing

$$f(x) = x^{p-1} + \cdots + x^2 + x + 1 = \frac{x^p - 1}{x - 1},$$

we make a change of variable to obtain $g(X) = f(X + 1)$. (From analytic geometry we recognize this as a translation of the $f(x)$ axis to a new $g(X)$ axis one unit to the left):

$$g(X) = f(X + 1) = \frac{(X + 1)^p - 1}{(X + 1) - 1}$$

$$= \left(X^p + pX^{p-1} + \frac{p(p - 1)}{2} X^{p-2} + \cdots + pX + 1 - 1 \right) \Big/ X$$

$$= X^{p-1} + pX^{p-2} + \frac{p(p - 1)}{2} X^{p-3} + \cdots + p.$$

Since it is known that all the binomial coefficients $\binom{p}{i}$, $i = 1, 2, \ldots, p - 1$, are divisible by p, we see that $g(X)$ satisfies the hypothesis of Theorem 6–7 and so is irreducible over $Q[x]$. Then $f(x)$ is irreducible over $Q[x]$, for if it factored into $f_1(x)f_2(x)$, we could form $g_1(X) = f_1(X + 1)$ and so factor $g(X)$. ∎

Definition 6–5. Let P and Q be polynomials in $F[x]$ which are not both zero. A polynomial D is a **greatest common divisor** or g.c.d. of P and Q if it is a common divisor ($D \mid P$ and $D \mid Q$) that is divisible by every common divisor ($E \mid P$ and $E \mid Q$ implies $E \mid D$). We write $D = (P, Q)$.

The following theorem is an algorithm or process that always enables us to find a greatest common divisor for two polynomials in $F[x]$ and to express that divisor linearly in terms of the given polynomials. As you will see, it is based on the division algorithm.

Theorem 6–8 (Euclidean Algorithm). Let P and Q be polynomials in $F[x]$ with $Q \neq 0$. There exist polynomials S and T in $F[x]$ such that

$$(P, Q) = SP + TQ.$$

PROOF: (Follow the example in Figure 6–7.) From the division algorithm there are quotient and remainder polynomials Q_1 and R_1 for which

$$P = Q_1Q + R_1,$$

$R_1 = 0$ or $\deg R_1 < \deg Q$; if $R_1 \neq 0$, apply the division algorithm to Q and R_1:

$$Q = Q_2R_1 + R_2,$$

$R_2 = 0$ or $\deg R_2 < \deg R_1$; if $R_2 \neq 0$, apply the division algorithm to R_1 and R_2:

$$R_1 = Q_3R_2 + R_3,$$

$R_3 = 0$ or $\deg R_3 < \deg R_2$; and so forth.

If at some stage the remainder R_i is the zero polynomial, then the process ends; we shall prove that R_{i-1} for $i > 1$, or Q in case $i = 1$, is the greatest common divisor. If no $R_i = 0$, then eventually the sequence $\deg Q > \deg R_1 > \deg R_2 > \cdots$ produces a remainder of degree zero, which we shall prove is (P, Q).

Take F to be the field of rational numbers, and take the initial polynomials

$$P = 6x^5 + 27x^4 - 23x^3 - 78x^2 - 57x - 14$$

and

$$Q = 3x^4 + 12x^3 - 19x^2 - 28x - 12.$$

The successive divisions give us:

$$P = Q_1Q + R_1$$

$$6x^5 + 27x^4 - 23x^3 - 78x^2 - 57x - 14$$
$$= (2x + 1)(3x^4 + 12x^3 - 19x^2 - 28x - 12) + (3x^3 - 3x^2 - 5x - 2)$$

$$Q = Q_2R_1 + R_2$$

$$3x^4 + 12x^3 - 19x^2 - 28x - 12$$
$$= (x + 5)(3x^3 - 3x^2 - 5x - 2) + (x^2 - x - 2)$$

$$R_1 = Q_3R_2 + R_3$$

$$3x^3 - 3x^2 - 5x - 2 = (3x)(x^2 - x - 2) + (x - 2)$$

$$R_2 = Q_4R_3 + R_4$$

$$x^2 - x - 2 = (x + 1)(x - 2) + 0.$$

Since $R_4 = 0$, R_3 is the greatest common divisor. To express it linearly in terms of P and Q, we substitute from the divisions in reverse order: $R_3 = R_1 - Q_3R_2$. $x - 2 = R_1 - Q_3R_2 = R_1 - Q_3(Q - Q_2R_1) = -Q_3Q + (1 + Q_2Q_3)R_1$ (by collecting terms) $= -Q_3Q + (1 + Q_2Q_3)(P - Q_1Q)$ (substituting for R_1) $= (1 + Q_2Q_3)P + (-Q_1 - Q_3 - Q_1Q_2Q_3)Q$ (collecting terms). Then $(P, Q) = SP + TQ$ with $S = 1 + Q_2Q_3 = 3x^2 + 15x + 1$, $T = -Q_1 - Q_3 - Q_1Q_2Q_3 = -6x^3 - 33x^2 - 20x - 1$.

FIGURE 6–7

Suppose $R_i = 0$. Then from

$$R_{i-2} = Q_iR_{i-1} + 0,$$

we have that $R_{i-1} \mid R_{i-2}$. From the preceding division

$$R_{i-3} = Q_{i-1}R_{i-2} + R_{i-1},$$

thus R_{i-1} divides the right member and therefore the left:

$$R_{i-1} \mid R_{i-3}.$$

Working back through the sequence of divisions, we can show similarly that R_{i-1} divides R_2 and R_1, so that

$$R_{i-1} \,|\, Q = Q_2 R_1 + R_2,$$

and hence

$$R_{i-1} \,|\, P = Q_1 Q + R_1.$$

In case $i = 1$, this sequence condenses to $R_1 = 0$, and $P = Q_1 Q$, so that $Q = (P, Q)$.

We have proved that R_{i-1} is a common divisor of P and Q if $R_i = 0$. Suppose there is not a zero remainder but eventually a remainder of degree zero, a nonzero field element. Such a field element is automatically a divisor of P and Q, whose coefficients are in the field.

Now since we can rewrite $R_{j-1} = Q_{j+1} R_j + R_{j+1}$ at any stage as $R_{j+1} = R_{j-1} - Q_{j+1} R_j$, we have a way of expressing each remainder in terms of earlier remainders in the sequence. In fact, by substituting for each remainder in turn, we can express R_{i-1}, if the sequence yields a remainder $R_i = 0$, or the remainder of degree zero otherwise, as a linear form

$$SP + TQ.$$

We already know that this polynomial is a common divisor of P and Q. From its form we now see that any common divisor of P and Q must divide $SP + TQ$. Hence $SP + TQ$ is the desired greatest common divisor

$$(P, Q) = SP + TQ. \quad \blacksquare$$

Exercise 6–29. Let a and b be positive integers. A division algorithm analogous to Theorem 6–3 gives

$$a = qb + r,$$

where q and r are integers and $0 \le r < b$. On this basis a Euclidean algorithm analogous to Theorem 6–8 enables us to express the greatest common divisor (a, b) of a and b linearly in terms of a and b:

$$(a, b) = sa + tb.$$

Find $(104, 65)$ linearly in terms of 104 and 65.

Exercise 6–30. Use Theorem 6–8 to prove that if polynomials P and Q have no factor of positive degree in common, then there are polynomials S and T for which $1 = SP + TQ$.

Theorem 6–9. The irreducible polynomials in $F[x]$ are prime polynomials in a sense analogous to Definition 5–1: If an irreducible polynomial P divides a product QR in $F[x]$, then P divides Q or P divides R.

PROOF: Suppose P does not divide Q. Then it has no polynomial of degree > 0 as a factor in common with Q, because P is irreducible. Then from Exercise 6–30 we have polynomials S and T for which $1 = SP + TQ$. Multiply both members of this equation by R to obtain

$$R = SPR + TQR.$$

By hypothesis P divides the last term, $T(QR)$. P also divides SPR. Then from $R = (SR + T(QR/P))P$, we see that P must divide R. ∎

Theorem 6–10. The factorization of a polynomial into irreducible polynomials in $F[x]$ is unique except for order and scalar multiples.

PROOF: Let $P_1 P_2 \cdots P_s$ and $Q_1 Q_2 \cdots Q_t$ be two factorizations of a polynomial in $F[x]$ into irreducible polynomials. From

$$P_1 P_2 \cdots P_s = Q_1 Q_2 \cdots Q_t$$

we see that P_1 divides the left member, hence the right member. From successive applications of Theorem 6–9, P_1 divides one of the polynomials Q_j, $j = 1, 2, \ldots, t$, for either P_1 divides Q_1 or it divides $Q_2 \cdots Q_t$, and if the latter, P_1 divides Q_2 or $Q_3 \cdots Q_t$, and so forth. Similarly, each P_i must divide some Q_j, and each Q_j must divide some P_i. By exhaustion, $s = t$, and any difference between the factorizations is in order or scalar multipliers from F. ∎

Theorem 6–10 has its counterpart for integers in the Fundamental Theorem of Arithmetic, which states that factorization of an integer into primes is unique except for order and unit factors.

Theorem 6–11. A polynomial equation $f(x) = 0$ with coefficients in a field F can have no more distinct roots in F than its degree.

PROOF: Let the degree of $f(x)$ be d and let r_1 be a root of the equation, that is, a zero of $f(x)$. Then, by Theorem 6–4, $x - r_1$ is a factor of $f(x)$, and we have

$$f(x) = (x - r_1)q_1(x);$$

comparing degrees on the two sides of the equation, we have

$$\deg q_1(x) = d - 1.$$

Now if $r_2 \neq r_1$ is also a root of the equation, we have from Theorem 6–4

$$(x - r_2) \,|\, f(x) = (x - r_1)q_1(x).$$

Since $(x - r_2) \nmid (x - r_1)$, we have by Theorem 6–9

$$(x - r_2) \,|\, q_1(x) \quad \text{or} \quad q_1(x) = (x - r_2)q_2(x)$$

with $\deg q_2(x) = d - 2$. If there are n distinct roots then

$$q_{n-1} = (x - r_n)q_n(x),$$

with $\deg q_n(x) = d - n$. But $q_n(x) \in F[x]$ has nonnegative degree, so $d - n \geq 0$, and $n \leq d$. ∎

Exercise 6–31. Refer to Exercise 6–22 and prove that the fifth degree equation studied there has exactly two complex roots.

CHAPTER 7

ALGEBRAIC FIELD
EXTENSIONS

We have used equivalence classes of number pairs to extend N to I. Then equations $x + a = 0$, $a \in N$, having no solution in N could be solved in I. Again, we used equivalence classes of number pairs to extend I to Q. Then equations $bx + a = 0$, $a, b \in I$, $b \neq 0$, having no solution in I could be solved in Q. Our object now is to form an extension E of Q that will contain a solution for a polynomial equation of finite degree n over Q

$$p(x) = p_0 + p_1 x + \cdots + p_n x^n = 0, \quad p_i \in Q \quad \text{for} \quad i = 0, 1, \ldots, n, \quad p_n \neq 0.$$

In fact, we can make our problem slightly more general by supposing that the coefficients come from a general field F. This will be useful, because we shall want to extend the extension. Once we have an extension E that contains one root r of $p(x) = 0$, we can factor $p(x)$ as $(x - r)q(x)$ in $E[x]$ and extend E to D if necessary to include a root of $q(x) = 0$, that is, a second root of $p(x) = 0$. Continuing the process we will eventually have an extension $\bar{E}$ containing all the roots of $p(x) = 0$. The nested algebras are shown schematically in Figure 7–1.

Definition 7–1. If every element of a field F is an element of a field E and if F has the same field operations as E, then F is a **subfield** of E, and E is an **extension** of F. We write $F \subseteq E$.

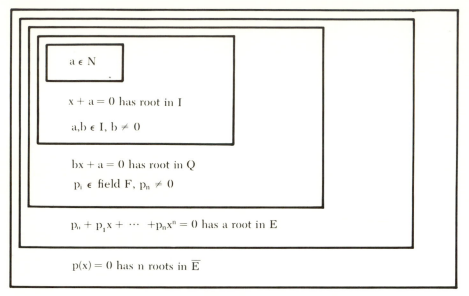

$a \in N$

$x + a = 0$ has root in I

$a, b \in I, \ b \neq 0$

$bx + a = 0$ has root in Q

$p_i \in$ field F, $p_n \neq 0$

$p_0 + p_1 x + \cdots + p_n x^n = 0$ has a root in E

$p(x) = 0$ has n roots in $\overline{E}$

FIGURE 7–1

Algebraic Numbers and Transcendental Numbers

We have considered the real numbers as an extension of the rational numbers, so that the set of real numbers can be thought of as the union of the set of rational numbers with the set of irrational numbers. The set of real numbers can also be thought of as the union of the real algebraic numbers and the real transcendental numbers.

Definition 7–2. Let $\sum_{i=0}^{n} a_i x^i = 0$ be a polynomial equation of degree n with coefficients c_i in a field F, and let E be an extension field of F. If $r \in E$ is a root of such an equation, then r is said to be **algebraic over F.** In case F is the field Q of rational numbers, r is an **algebraic number.**

Definition 7–3. A real number that is not the root of any polynomial equation with rational coefficients is called **transcendental.**

The two roots of a quadratic equation $ax^2 + bx + c = 0$ in $R[x]$, $a \neq 0$,

$$\frac{(-b + \sqrt{b^2 - 4ac})}{2a} \quad \text{and} \quad \frac{(-b - \sqrt{b^2 - 4ac})}{2a},$$

are algebraic numbers, for they are roots of the polynomial equation of degree 2 over R.

Definition 7–4. An algebraic number is called an **algebraic integer** if it is a root of a polynomial equation of degree n with coefficients in I with the coefficient of x^n equal to 1.

Notice that every algebraic number is a root of an equation with coefficients in I, for from Definition 7–2 it is a root of an equation with coefficients in Q. Multiply the equation by the least common multiple of the denominators of the rational coefficients. The result is an equation with integral coefficients, the coefficient of x^n not necessarily 1.

The integers I are algebraic integers, for each $b \in I$ satisfies the equation $x - b = 0$ of degree 1, coefficients in I, leading coefficient 1. The elements of I are sometimes called the "rational integers," distinguishing them from the other algebraic integers.

Classroom Exercise 7–1. Show that $(7 + \sqrt{61})/2$ is an algebraic integer. (Form an equation having this as one root, and $(7 - \sqrt{61})/2$ as another.)

How many algebraic numbers are there? In Appendix A we use the positive integers N written to base 13 to count the equations $bx = a$. A judicious extension of this technique provides a one-to-one correspondence between the elements of N and the polynomial equations with integral coefficients, proving that there are only $\aleph_0$ algebraic numbers.

Now, bringing together the facts that there are only $\aleph_0$ algebraic numbers but $2^{\aleph_0}$ real numbers and that all the nonalgebraic real numbers are transcendental, one can show that the transcendental ones are not countable.

If in R the transcendental numbers are not countable, while the algebraic numbers are countable, you would expect to encounter transcendental numbers much more frequently than algebraic ones. As a matter of fact, only a few decimal numbers known to be transcendental are important enough to have turned up in your previous mathematical experience, notably π and e. The first enters early in formulas connected with circles, for it relates the circumference c of a circle to the radius r:

$$c = 2\pi r.$$

The area A of the circle is

$$A = \pi r^2.$$

In some problems the approximation 22/7 is a close enough value for π, 22/7 being the root of the first degree equation $7x = 22$. Sometimes the approximations 3, 3.14, or 3.1416 are used, each of these also the root of a first-degree equation. One state legislature even undertook to make life easier for students by setting the value of π at exactly 3 by legislative act! However, circles remained the same outside politics. It can be shown that π, far from being rational, as it would be if its decimal expansion terminated (Theorem 2–6), is not even the root of a quadratic, third-degree, nor indeed any finite degree equation. An infinite series expansion of $\pi/4$ gives us

$$\frac{\pi}{4} = 1 - \frac{1}{3} + \frac{1}{5} - \frac{1}{7} + \frac{1}{9} - \cdots.$$

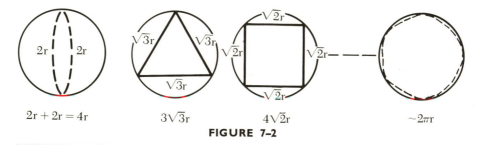

$$2r + 2r = 4r \qquad\qquad 3\sqrt{3}r \qquad\qquad 4\sqrt{2}r \qquad\qquad \sim 2\pi r$$

FIGURE 7–2

The first proof that π is transcendental was made in 1882 by Ferdinand Lindemann,† but clever new proofs still appear in the literature.

The constant e (named for Euler), much used in calculus as the base of natural logarithms, can also be shown to be transcendental. It was first proved transcendental by Hermite in 1873.† It can be represented by the infinite series

$$e = 1 + \frac{1}{1!} + \frac{1}{2!} + \frac{1}{3!} + \cdots + \frac{1}{n!} + \cdots.$$

There are constructions for other transcendental numbers, in fact, whole classes of them, but their main interest for us is the fact that they exist.

---

† See, for example, R. L. Wilder, *Introduction to the Foundations of Mathematics*, Wiley, 1958, p.88, and his reference to Sierpinski.

> **Theorem 7–1.**   Let F be a field, and let $\theta \in E \supseteq F$ be algebraic relative to F. Then θ is a zero of a unique monic irreducible polynomial $p(x) \in F[x]$, and $p(x)$ divides each polynomial in $F[x]$ that has θ as a zero.

PROOF: From the definition of an algebraic extension E, each $\theta \in E$ is a zero of some polynomial in $F[x]$.

Let $p(x)$ be a monic polynomial of minimal degree in $F[x]$ having θ as a zero. (Since F is a field, a polynomial whose leading coefficient is $c \neq 0$ can be made monic by multiplication throughout by c^{-1}.)

First, note that $p(x)$ is irreducible, for if $p(x) = p_1(x)p_2(x)$, where both factors have positive degrees, we have $p(\theta) = p_1(\theta)p_2(\theta) = 0$, so that θ is a zero of $p_1(x)$ or of $p_2(x)$, each of which has degree less than the degree of $p(x)$.

Now let $g(x)$ be any polynomial in $F[x]$ for which $g(\theta) = 0$. We can prove that $p(x) \mid g(x)$, for by the division algorithm (Theorem 6–3)

$$g(x) = p(x)q(x) + r(x),$$

where the quotient $q(x)$ and the remainder $r(x)$ are in $F[x]$ and $r(x)$ is the zero polynomial or $\deg r(x) < \deg p(x)$. Again using the homomorphism of Theorem 6–2 we have

$$0 = g(\theta) = p(\theta)q(\theta) + r(\theta),$$

so that

$$0 = 0 \cdot q(\theta) + r(\theta),$$

implying that $r(\theta) = 0$. Then θ is a zero of some irreducible factor of $r(x)$. But $p(x)$ was of minimal degree among irreducible polynomials with θ as a zero, so $r(x)$ is the zero polynomial, and $p(x) \mid g(x)$. In particular, if $g(x)$ is monic and irreducible, $p(x) = g(x)$, so that $p(x)$ is unique.   ∎

> **Definition 7–5.**   Let F be a field. Then θ_1 and $\theta_2 \in E \supseteq F$ are **conjugates** if they are zeros of the same irreducible polynomial in $F[x]$.

Exercise 7–2.   The complex numbers $s + ti$ and $s - ti$, $s, t \in R$, are often called conjugates. Show that they satisfy Definition 7–5. What happens if $t = 0$?

Definition 7–6. If every element of an extension E of F is algebraic relative to F, then E is an **algebraic extension** of F.

Definition 7–7. Let F be a field, and let $p(x)$ be a polynomial with coefficients in F irreducible in $F[x]$. We let **$F[x]/\langle p(x) \rangle$** stand for the polynomials with coefficients in F taken modulo $\langle p(x) \rangle$, with members the equivalence classes $\{r(x) + p(x)q(x), \text{ where } q(x) \in F[x]\}$, represented by $r(x) \bmod \langle p(x) \rangle$.

Each polynomial $f(x)$ is equivalent modulo $\langle p(x) \rangle$ to its remainder $r(x)$ upon division by $p(x)$. From the Division Algorithm (Theorem 6–3) the remainder is uniquely determined as 0 or as a polynomial of degree less than the degree of $p(x)$. (See Figure 7–3.)

Exercise 7–3. Let F be the field of rational numbers and let $p(x) = x^2 + x/3 - 1$. Show that $x^3 + 1$, $-x^2/3 + x + 1$, and $10x/9 + \frac{2}{3}$ are in the same equivalence class. Show that $x^4 + x^2 + 1$ and $-28x/27 + \frac{28}{9}$ are in the same equivalence class.

As an example of representing polynomials modulo $\langle p(x) \rangle$, take $p(x) = x^2 + 1$, which is irreducible over the rationals $Q[x]$. To represent the polynomial $3x^5 - 5x^3 + 2x + 1$ modulo $\langle x^2 + 1 \rangle$, we divide the polynomial by $x^2 + 1$, obtaining the remainder $10x + 1$. We may write the member of $F[x]/\langle p(x) \rangle$ with a bar over it, $\overline{10x + 1}$, to show that it stands for the entire equivalence class $\{10x + 1 + (x^2 + 1)q(x)\}$.

$$
\begin{array}{r}
3x^3 \; -8x \\
x^2 + 1 \overline{)3x^5 \; -5x^3 \; +2x \; +1} \\
\underline{3x^5 \; +3x^3 } \\
-8x^3 \; +2x \; +1 \\
\underline{-8x^3 \; -8x } \\
10x \; +1
\end{array}
$$

$$3x^5 - 5x^3 + 2x + 1 \to \overline{10x + 1}$$
$$\text{Any } 10x + 1 + (x^2 + 1)q(x) \to \overline{10x + 1}$$

FIGURE 7–3

Take $p(x) = x^2 + 1$, which is irreducible in $Q[x]$. Let

$$f = x^3 + 7x^2 - 9, \qquad g = 2x^3 + 7,$$
$$\bar{r} = -x - 16, \qquad \bar{s} = -2x + 7.$$

$\phi(f + g) =$
$$\phi(3x^3 + 7x^2 - 2) = (-3x - 9)(\mathrm{mod}\langle x^2 + 1\rangle) = \bar{r} + \bar{s}(\mathrm{mod}\, p)$$

$\phi(f \cdot g) =$
$$\phi(2x^6 + 14x^5 - 11x^3 + 49x^2 - 63) = (25x - 114)(\mathrm{mod}\langle x^2 + 1\rangle).$$

$\phi(f) \cdot \phi(g) =$
$$(\bar{r} \cdot \bar{s})(\mathrm{mod}\, p) = (2x^2 + 25x - 112)(\mathrm{mod}\, p) \equiv (25x - 114)(\mathrm{mod}\, p).$$

FIGURE 7-4

Classroom Exercise 7–4.   Continuing Exercise 7–3, find linear polynomials equivalent to x^2, to x^3, and to x^4. Use the results to find a linear polynomial equivalent to $3x^4 + x^3 - 2x^2/3 + x + 1$.

Theorem 7–2 (Kronecker).   Let F be a field, and let $p(x)$ be irreducible in $F[x]$. Then $F[x]/\langle p(x)\rangle$ is a field. It contains a subfield isomorphic to F. The polynomial $p(x)$ has a zero in $F[x]/\langle p(x)\rangle$.

PROOF: The arithmetic of $F[x]$ suggests an arithmetic for $F[x]/\langle p(x)\rangle$: For the sum of two equivalence classes $\bar{r} = r(\mathrm{mod}\, p)$ and $\bar{s} = s(\mathrm{mod}\, p)$, we use the equivalence class containing $\bar{r} + \bar{s}$; for the product, the equivalence class containing $\bar{r} \cdot \bar{s}$. (See Exercise 7–5.)

Classroom Exercise 7–5.   Show that the sum and product just defined are not dependent on which representatives are selected from the equivalence classes: If f and r are in the same equivalence class, then $f = r + pq$, for $q \in F[x]$, by Definition 7–7. Compare $r + s$ and $f + g$, with $g = s + pq_2$. Compare $r \cdot s$ and $f \cdot g$.

Let ϕ map each polynomial $f(x)$ to its equivalence class $\{r(x) + p(x)q(x)\}$ modulo $p(x)$. Then ϕ is a homomorphism from $F[x]$ to $F[x]/\langle p(x)\rangle$, preserving $+$ and $\cdot$, as shown in Classroom Exercise 7–5 and illustrated in Figure 7–4. The equivalence classes form a ring with unit element with $+$ and $\cdot$ as defined in the proof.

Exercise 7–6.   Prove that $+$ and $\cdot$ are binary operations on $F[x]/\langle p(x)\rangle$.

Exercise 7–7.   What is the multiplicative identity in $F[x]/\langle p(x)\rangle$?

To show that $F[x]/\langle p(x)\rangle$ is a field, we need to show that for any $\overline{r(x)} \not\equiv 0(\mathrm{mod}\langle p(x)\rangle)$ there is an inverse $\overline{s(x)}$ for which $\overline{s(x)} \cdot \overline{r(x)} \equiv \overline{1}(\mathrm{mod}\langle p(x)\rangle)$. It was mainly for this purpose that we introduced the Euclidean Algorithm (Theorem 6–8) and Exercise 6–30. Since $p(x)$ is irreducible in $F[x]$, it has no factor of positive degree except scalar multiples of itself and, hence, no factor of positive degree in common with $r(x) \not\equiv 0$. Then the greatest common divisor of $r(x)$ and $p(x)$ is a field element d, which can then be expressed linearly in terms of $r(x)$ and $p(x)$. Multiplying by d^{-1} we can express the multiplicative identity 1 of F as

$$1 = s(x)r(x) + t(x)p(x).$$

In $F[x]/\langle p(x)\rangle$, the last term is in the $\overline{0}$ equivalence class, so we have

$$1 \equiv \overline{s(x)} \cdot \overline{r(x)} \quad (\mathrm{mod}\langle p(x)\rangle),$$

and $\overline{s(x)}$ is the required inverse. (Computation of an inverse is illustrated in Figure 7–5.)

Exercise 7–8.   Check the inverse computed in Figure 7–5 by multiplication in $Q[x]/\langle x^2 + 1\rangle$.

Exercise 7–9.   In Exercise 6–8 we proved that in $F[x]$ there is no inverse for the polynomial x. Find the inverse of the equivalence class $\bar{x}$ in

$$Q[x]/\langle x^2 + 1\rangle.$$

The field $F[x]/\langle p(x)\rangle$ has the field F as a subfield up to an isomorphism, for every field element, or constant, d, is its own representative $\bar{d}$ and operations in $F[x]/\langle p(x)\rangle$ reduce to the operations in F when applied to constants, for no reduction modulo $\langle p(x)\rangle$ is necessary.

Now we can show that in $F[x]/\langle p(x)\rangle$ the polynomial function $p(x)$ has $\bar{x}$ as a zero. Notice that in $F[x]/\langle p(x)\rangle$ $\bar{x}$ is not an indeterminate but is one of the equivalence classes making up the field. In case the degree of $p(x)$ is 1, the linear polynomial has a zero in F and, hence, certainly, in $F[x]/\langle p(x)\rangle$. If $p(x) = p_0 + p_1x + p_2x^2 + \cdots + p_nx^n$, then since the reduction of $F[x]$ modulo $\langle p(x)\rangle$ is a homomorphism with respect to $+$ and $\cdot$, we have

$$\overline{p(x)} \equiv \bar{p}_0 + \bar{p}_1\bar{x} + \bar{p}_2\bar{x}^2 + \cdots + \bar{p}_n\bar{x}^n.$$

But in $F[x]/\langle p(x)\rangle$ we have $\overline{p(x)} \equiv \overline{0}$, so that

$$\bar{p}_0 + \bar{p}_1\bar{x} + \bar{p}_2\bar{x}^2 + \cdots + \bar{p}_n\bar{x}^n \equiv 0$$

and x is a zero of $\bar{p}(x)$. As we have shown, F is isomorphic to a subfield of $F[x]/\langle p(x)\rangle$. Under this isomorphism $p(x)$ corresponds to $\bar{p}(x)$, justifying the assertion of the theorem that $p(x)$ has a zero in $F[x]/\langle p(x)\rangle$. ∎

Find $\overline{(x + \frac{1}{2})}^{-1}$ in $Q[x]/\langle x^2 + 1\rangle$. We use the Euclidean algorithm (Theorem 6–8) to express the g.c.d. of $x + \frac{1}{2}$ and $x^2 + 1$.

$$
\begin{array}{r}
x - \dfrac{1}{2} \\[4pt]
x + \dfrac{1}{2}\overline{\smash{)}\,x^2 + 1} \\[4pt]
x^2 + \dfrac{x}{2} \\[2pt]
\hline
-\dfrac{x}{2} + 1 \\[4pt]
-\dfrac{x}{2} - \dfrac{1}{4} \\[2pt]
\hline
\dfrac{5}{4}
\end{array}
$$

$$
\frac{5}{4} = x^2 + 1 - \left(x - \frac{1}{2}\right)\left(x + \frac{1}{2}\right).
$$

Then

$$
1 = \left(\frac{4}{5}\right)\left(\frac{5}{4}\right) = \left(\frac{4}{5}\right)\left[x^2 + 1 - \left(x - \frac{1}{2}\right)\left(x + \frac{1}{2}\right)\right],
$$

$$
\bar{1} \equiv -\left(\frac{4}{5}\right)\overline{\left(x - \frac{1}{2}\right)} \cdot \overline{\left(x + \frac{1}{2}\right)}
$$

$$
\equiv \overline{\left(\frac{-4x}{5} + \frac{2}{5}\right)} \cdot \overline{\left(x + \frac{1}{2}\right)}.
$$

The inverse of $\overline{x + \frac{1}{2}}$ is $\overline{-4x/5 + \frac{2}{5}}$, as can be checked by multiplication in $Q[x]/\langle x^2 + 1\rangle$.

FIGURE 7–5

Now that we have established that $p(x)$ has a zero in the extended field $F[x]/\langle p(x)\rangle$, let θ stand for a zero of $p(x)$. Then every element of the extended field $F[x]/\langle p(x)\rangle$ can be represented in the form $r(\theta)$, where $r(x)$ is a polynomial in $F[x]$ reduced modulo $\langle p(x)\rangle$. Then by the Factor Theorem 6–4 $x - \theta$ is a factor of $p(x)$ in the extended field.

Theorem 7–2 deserves some comment. In Chapter 2 we extended the natural number system N to obtain all the integers I by forming an equivalence class of problems and calling that class the *answer* to the problems it contained. We used the same approach to enlarge the system of integers I to obtain the rational field Q.

In the present case our problem is to find a zero for a polynomial $p(x)$ that has none in F. In essence our approach is to form classes of polynomials in θ that are alike provided θ is treated as a zero of $p(x)$. We select a representative for each such class and provide for reducing sums and products using the relation $p(\theta) = 0$. Then we prove that the particular element θ is a zero of $p(x)$, as indeed it was constructed to be!

As we have emphasized previously, this solution of a problem by equating the problem to the solution is typical of algebra. It appears in elementary algebra when we ask "How old is Mary if she is twice as old as a $15\frac{1}{2}$-year-old boy?" We let x stand for Mary's age, and say that x is the answer. But that was the problem! All we have done is give the problem the name x and hand it back as its own solution. To be scrupulously fair we look around for the widest range of problems for which this x can serve as name and solution, obtaining the equivalence class of problems

$$x = \{2(15\tfrac{1}{2}) = 31\},$$

in this case a 1-membered class.

Notice the way the proof of Theorem 7–2 shows how to find the inverse of any $r(\theta) \not\equiv 0$. A familiar way to find the inverse of the complex number $s + ti$ is to write

$$\frac{1}{s + ti} = \frac{1}{s + ti} \cdot \frac{s - ti}{s - ti} = \frac{s - ti}{s^2 + t^2}.$$

According to the proof of Theorem 7–2, the trick multiplication by $(s - ti)/(s - ti)$ can be replaced by a computation similar to that of Figure 7–5, and this computation generalizes to irreducible polynomials other than $x^2 + 1$.

Exercise 7–10. Following Figure 7–5, divide $\theta^2 + 1$ by $t\theta + s$, express their g.c.d. linearly, find the inverse of $s + ti$, and compare with the preceding result.

Exercise 7–11. Suppose that θ is a zero of $p(x) = x^5 - 4x^4 + 2x + 2$ in $Q[x]/\langle p(x) \rangle$ (cf. Exercises 6–22, 6–28, 6–31). Illustrate the homomorphism ϕ of Theorem 7–2 by finding the image of the sum and product and the sum and product of the images for $f = x^6 - 2x^4$ and $g = x^5$.

Exercise 7–12. Continuing the previous exercise, find the inverse of θ in the extension field. (Notice that $\theta^5 - 4\theta^4 + 2\theta + 2 = \theta(\theta^4 - 4\theta^3 + 2) + 2$.)

Exercise 7–13. Continuing the previous exercises, show by synthetic division that $x - \theta$ is a factor of $x^5 - 4x^4 + 2x + 2$ in the extension field. What is the quotient polynomial?

Exercise 7–14. In Exercise 7–12 a way of finding θ^{-1} is suggested based on the constant, or zero-degree term, p_0 of $p(x)$. Prove that if $p(x)$ is irreducible in $F[x]$, then its constant term p_0 is not zero in F. From $p(\theta) \equiv 0$, write $-p_0$ in $F[x]/\langle p(x) \rangle$ linearly in powers of θ. Factor out θ, divide by $-p_0$, and so find θ^{-1} as in Exercise 7–12. Check by finding i^{-1} in $Q[x]/\langle x^2 + 1 \rangle$.

By the congruence-class construction of Theorem 7–2, we have reduced the irreducible polynomial $p(x)$; we have constructed an extension field E in which $p(x)$ has at least one zero θ. Now we prepare to extend E to include more zeros of $p(x)$.

Definition 7–8. Let D be an extension of a field F and let $\theta_1, \theta_2, \ldots, \theta_k$ be elements of D. Then $E = F(\boldsymbol{\theta}_1, \boldsymbol{\theta}_2, \ldots, \boldsymbol{\theta}_k)$ (read "F with $\boldsymbol{\theta}_1, \boldsymbol{\theta}_2, \ldots, \boldsymbol{\theta}_k$ adjoined") is defined to have as elements the intersection:
$\cap$ {the subfields E' for which $F \subseteq E' \subseteq D$ and $\theta_1, \theta_2, \ldots, \theta_k \in E'$}; that is, the elements of E are the elements common to all subfields E' of D that contain F and the θ's. As we shall prove, E is a field. (See diagram in Figure 7–6.)

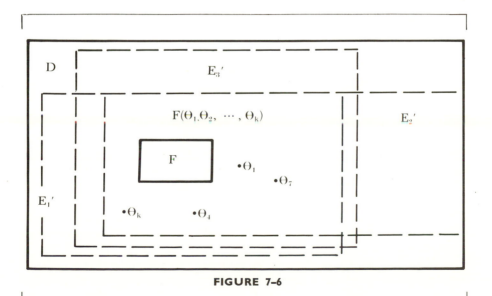

FIGURE 7–6

Lemma 7–1. Let J be a set of subgroups of a group. The intersection $M = \cap\{G \mid G \in J\}$ of subgroups in J is a group. If all the subgroups in J are abelian, then M is abelian.

PROOF: Each element of M is an element of the subgroups G, hence of a group, so we apply Theorem 3–6 to prove that M is a subgroup. If m and n are elements of M, then they are elements of every subgroup G. Then $m * n$, where $*$ is the group operation, is an element of every G, and therefore of M. Then M is closed under $*$. If $m \in M$, then $m \in G$ for each G. Then $m^{-1} \in G$ for each G, so that $m^{-1} \in M$. Then M is a subgroup. If $m * n = n * m$ in all the subgroups G, then $m * n = n * m$ in M, so that M is abelian. ∎

Theorem 7–3. Let F be a field. Let D be an extension of F that contains $\theta_1, \theta_2, \ldots, \theta_k$. Then

i. $E = F(\theta_1, \theta_2, \ldots, \theta_k)$ is a subfield of D and an extension of F.

ii. The elements of E are the rational combinations of the θ's and the elements of F, where a combination is rational if it is formed from quotients of sums and products, zero denominators excluded.

iii. If $\theta \in D$ is a zero of $p(x)$ of degree n irreducible in $F[x]$, then $F(\theta) \cong F[x]/\langle p(x) \rangle$. Each element of $F(\theta)$ can be expressed uniquely as a linear combination of 1, θ, $\theta^2, \ldots, \theta^{n-1}$ with coefficients in F.

PROOF: This theorem is a so-called "housekeeping" theorem; it puts together the details we need for actually using the field adjunctions of Definition 7–8. We illustrate its various provisions in Figure 7–7.

We proceed with the proof itself. To prove *i* we first note that there exist fields E' with $F \subseteq E' \subseteq D$ and $\theta_1, \theta_2, \ldots, \theta_k \in E'$, for D is such a field. The intersection E of fields E' is a subfield of D, for, by Lemma 7–1, the intersection forms an abelian group under addition, at least the two elements 0 and 1 occur in every E', hence in E, multiplication is associative in E, and, again from Lemma 7–1, the nonzero elements of E form an abelian group under multiplication. The distributive property is inherited. Since each E' is intermediate between F and D, so is the intersection.

To prove *ii* we notice first that each rational combination of the θ's has a form S/T, $T \neq 0$, where S and T are sums of products of the θ's and the elements of F, and so each S/T is an element of D. We can combine quotients

Let F stand for Q, the rational field, and let $\sqrt{-3}, \sqrt{5}$ be $\theta_1, \theta_2 \in C$, the complex field. Then Theorem 7–3 shows

i. $Q \subseteq Q(\sqrt{-3}, \sqrt{5}) \subseteq C$.

ii. $Q(\sqrt{-3}, \sqrt{5})$ is made up of members S/T, $T \neq 0$, where S and T are sums of products of rationals, $\sqrt{-3}$, and $\sqrt{5}$.

iii. Since $\sqrt{-3}$ is a zero of $x^2 + 3$ of degree 2 in $Q[x]$, $Q(\sqrt{-3}) \cong Q[x]/\langle x^2 + 3 \rangle$ and each element of $Q(\sqrt{-3})$ can be expressed uniquely in the form

$$a + b\sqrt{-3}, \quad a, b \in Q.$$

FIGURE 7–7

according to the arithmetic of D

$$\frac{S_1}{T_1} + \frac{S_2}{T_2} = \frac{(S_1 T_2 + S_2 T_1)}{T_1 T_2}$$

and

$$\left(\frac{S_1}{T_1}\right)\left(\frac{S_2}{T_2}\right) = \frac{S_1 S_2}{T_1 T_2}.$$

Then the rational combinations form a field $\bar{F}$, for they are closed under $+$ and $\cdot$ and the other requirements of Definition 2–5 can be verified for quotients S/T. Also, all elements of F and $\theta_1, \theta_2, \ldots, \theta_k$ are rational combinations of themselves and so are included in $\bar{F}$. We conclude that $\bar{F}$ is a subfield for which $F \subseteq \bar{F} \subseteq D$ and the θ's are in $\bar{F}$, so that $\bar{F}$ is one of the E's in Definition 7–8. Then the intersection $F(\theta_1, \theta_2, \ldots, \theta_k) \subseteq \bar{F}$. But every element of $\bar{F}$ is an element of $F(\theta_1, \theta_2, \ldots, \theta_k)$, for any S/T in $\bar{F}$ is a rational combination of the θ's and the elements of F, so $S/T \subseteq E'$ for every E' of Definition 7–8, proving that $S/T \subseteq \cap E'$. Then $\bar{F} \subseteq F(\theta_1, \theta_2, \ldots, \theta_k)$. Thus $\bar{F} = F(\theta_1, \theta_2, \ldots, \theta_k)$.

To prove iii, note that each rational combination of F-elements and θ can be expressed as the quotient $n(\theta)/d(\theta)$ of two polynomials in θ with coefficients in F and degrees less than the degree of $p(x)$ for $p(\theta) = 0$. The denominator $d(\theta) \neq 0$. Since $p(x)$ is irreducible, it has no factor of positive degree less than n, the degree of $p(x)$. Then there are polynomials $s(x)$ and $t(x)$ for which

$$1 = s(x)d(x) + t(x)p(x),$$

so that

$$1 = s(\theta)d(\theta) + t(\theta) \cdot 0.$$

Then $[d(\theta)]^{-1} = s(\theta)$.

Now form the product $n(\theta)s(\theta) = n(\theta)/d(\theta)$, and reduce it to a degree less than n by using the relation $p(\theta) = 0$, obtaining for the rational combination the unique form

$$f_0 + f_1\theta + f_2\theta^2 + \cdots + f_{n-1}\theta^{n-1}, \quad f_i \in F.$$

The correspondence $f(\theta) \leftrightarrow f(x)$ is an isomorphism with respect to $+$ and $\cdot$ between $F(\theta)$ and $F[x]/\langle p(x) \rangle$. From polynomials of degree zero, we note that each element of F corresponds to itself under the isomorphism. ∎

We see from this theorem that the field of equivalence classes modulo an irreducible $p(x) \in F[x]$ set up in Theorem 7–2 is the same as the field F together with all rational combinations of elements of F and θ, where θ is a zero of $p(x)$ in an extension field. In the case of the extension from the real field to the complex field, we took the point of view that we knew $x^2 + 1$ had a zero i in C, so we introduced all rational combinations of reals and i. Galois simply assumed there was a zero for any irreducible polynomial and "adjoined" it, that is, included all rational combinations. It was Cauchy who was led by the success of Galois toward the sophisticated method of Theorem 7–2.

Definition 7–9. Let E be an extension of F. Elements $r_1, r_2, \ldots, r_n$ of E form a **linear basis for E over F** if every element g of E can be expressed uniquely in the form $g = f_1 r_1 + f_2 r_2 + \cdots + f_n r_n$, with coefficients f_i in F. Elements $t_1, t_2, \ldots, t_j$ of $E \cdot$ are **linearly dependent** over F if there are coefficients $f_1, f_2, \ldots, f_j \in F$ not all zero for which $f_1 t_1 + f_2 t_2 + \cdots + f_j t_j = 0$; if there are no such coefficients, that is, if $f_1 t_1 + f_2 t_2 + \cdots + f_j t_j = 0$ implies all the coefficients f_i are zero, then the elements t_i are **linearly independent**. The n basis elements are linearly independent, for since 0 has the unique expression $0 = 0 \cdot r_1 + 0 \cdot r_2 + \cdots + 0 \cdot r_n$, a linear relation $f_1 r_1 + f_2 r_2 + \cdots + f_n r_n = 0$ would imply that all the coefficients are 0. If E has a linear basis of n elements over F, then the **degree $[E:F]$ of E over F** is n, and E is a **finite extension** of F. (For more material on linear independence, see any book on linear algebra or matrices, such as I. N. Herstein, *Topics in Algebra*, Blaisdell, 1964, Chapter 4.)

It can be shown by the technique of eliminating variables among m simultaneous linear equations in n variables that if $m > n$, then a set $s_1, s_2, \ldots, s_m$ of m elements of E cannot form a linear basis for E over F, because they are linearly dependent. (See illustration in Figure 7–8.) No basis of fewer than n elements can be found, either, for then a linear dependence could be found among the r_i's. Then the degree n of E over F is not ambiguously defined, as all linear bases contain the same number n of elements.

Theorem 7–4. Let F be a field. If $\theta \in D \supseteq F$ is a zero of $p(x)$ of degree n irreducible in $F[x]$, then $1, \theta, \theta^2, \ldots, \theta^{n-1}$ form a linear basis for $F(\theta)$ over F and $F(\theta)$ is an algebraic extension of F.

Let $n = 2$, and let r_1, r_2 be a linear basis for E over F. Take $m = 3$, and suppose

$$\begin{cases} s_1 = 2r_1 - r_2 \\ s_2 = 5r_1 + 3r_2 \\ s_3 = -r_1 + r_2 \end{cases}$$

Then we can eliminate r_2 among the three equations:

$$3s_1 + s_2 = 11r_1$$

$$s_1 + s_3 = r_1$$

from which

$$3s_1 + s_2 - 11(s_1 + s_3) = 0,$$

or $-8s_1 + s_2 - 11s_3 = 0$, with $-8, +1, -11 \neq 0$.

FIGURE 7–8

PROOF: Theorem 7–3iii shows that $1, \theta, \theta^2, \ldots, \theta^{n-1}$ satisfy the requirements of Definition 7–9 for a linear basis.

Let g be any nonzero element of $F(\theta)$. Since $1, g, g^2, \ldots, g^n$ can each be expressed linearly in terms of the n basis elements, they have a linear dependence

$$a_0 \cdot 1 + a_1 g + a_2 g^2 + \cdots + a_n g^n = 0.$$

Then g is a zero of $a_0 + a_1 x + a_2 x^2 + \cdots + a_n x^n \in F[x]$. ∎

Exercise 7–15.   Prove that $1, i = \sqrt{-1}$ form a linear basis for $R(i)$, where R is the real field. Show that $g = \sqrt{2} - 2i$ is algebraic relative to R by finding a polynomial having g as a zero.

As we have noted (see Exercise 5–14, page 95), there are fields with only a finite number of elements, such as the integers taken modulo 2 or modulo 5. However, several of our key theorems and Abel's theorem itself apply only to the infinite fields Q, R, C, and their extensions, which leads to the following definition.

Definition 7–10.   A field F is called a field of **characteristic zero** if F contains a subfield isomorphic to the field Q of rational numbers.

Lemma 7–2. Let F be any field and let $f(x)$ be a polynomial of degree d in $F[x]$. Then $f(x)$ has no more than d zeros in any extension field $E \supseteq F$.

Let F be a field of characteristic zero, and let $p(x)$ of degree n be a monic irreducible polynomial in $F[x]$. If $(x - \theta)^m$, with $m > 0$, divides $p(x)$ in $E[x]$, where E is an extension field of F, then m equals 1, so that the zeros of $p(x)$ in E are distinct.

PROOF: See Appendix I.

Definition 7–11. Let $F \subseteq E \subseteq D$ be fields. A polynomial $f(x) \in F[x]$ **splits** in D if it factors into linear factors in $D[x]$. D is a **splitting field** for $f(x)$ over F if $f(x)$ splits in $D[x]$ and if it does not split in E unless $E = D$.

Theorem 7–5. Let $f(x)$ of degree k in $F[x]$ have zeros θ_1, $\theta_2, \ldots, \theta_k$ in an extension D of F. Then $E = F(\theta_1, \theta_2, \ldots, \theta_k)$ is a splitting field for $f(x)$ over F. Every polynomial $f(x) \in F[x]$ of positive degree has a splitting field over F.

PROOF: Certainly $f(x)$ splits in $E = F(\theta_1, \theta_2, \ldots, \theta_k)$, for it has linear factors $x - \theta_i, i = 1, 2, \ldots, k$. (It may also have a constant factor $c \in F$, but it may have no more than k linear factors by Lemma 7.2.) Has E the minimal property required for a splitting field? Any subfield E' of D containing F in which $f(x)$ splits must contain the θ's and so must be one of the subfields E' of E used in Definition 7–6. Then E, as the intersection of all such fields E', is minimal.

Let $f(x)$ have a factor $p(x)$ irreducible in $F[x]$ of degree $n > 1$ and let F have characteristic zero. Then by the technique of Theorem 7–2 we can construct an extension E of F in which $p(x)$ has a zero θ_1. Then in $E[x], f(x)$ has a factorization $(x - \theta_1)q(x)$. If $q(x)$ has irreducible factors of degree greater than one in $E[x]$, we can use the construction of Theorem 7–2 to extend E to E_2 in which $q(x)$, hence $f(x)$, factors further.

By Lemma 7–2 $p(x)$ has no more zeros than its degree. Then by a finite number of successive extensions of F we can find an extension in which $f(x)$

splits. The hypotheses of the first part of this theorem are then satisfied, and $F(\theta_1, \theta_2, \ldots, \theta_k)$ is a splitting field for $f(x)$ over F. ∎

Exercise 7–16.   Construct a splitting field for the polynomial $x^4 - 8x^2 + 15$ over $Q[x]$.

At this point we have enough field theory to appreciate Galois theory, so we postpone to Appendix I proofs of some further results used in a proof of Abel's Theorem. We can describe these results in a general way: It turns out (Theorem I–1) that the order in which we adjoin $\theta_1, \theta_2, \ldots, \theta_k$ one at a time to a field F makes no difference, up to an isomorphism. Because of this, we can show (Theorem I–2) that the splitting field for a polynomial $f(x) \in F[x]$ over F is unique up to an isomorphism.

Theorem I–3 shows us how successive finite extensions of a field change the degree of the extension. If E is an extension of degree n over F and D in turn has degree m over E, then D has degree mn over F.

Theorem I–4 shows that k successive finite extensions of a field F of characteristic zero can always be accomplished by a single finite extension. We see incidentally that any finite extension is an algebraic extension.

Classroom Exercise 7–17.   Suppose $f(x) \in Q[x]$ has a complex zero $s + ti$, $t \neq 0$. Then it also has a complex zero $s - ti$. Prove this result by showing what happens to real and pure imaginary parts of $f(s + ti)$ and $f(s - ti)$.

Classroom Exercise 7–18.   Prove the same result as in Classroom Exercise 7–17 by this line of reasoning: First find a monic irreducible quadratic polynomial having $s + ti$ as a zero. Argue from Theorem 7–1 that the quadratic polynomial divides $f(x)$. Then use the Factor Theorem 6–4 to prove that $s - ti$ is a zero of $f(x)$.

Classroom Exercise 7–19.   Compare the two preceding exercises. The first method typifies the theory-of-equations approach that was once very popular. It gives you lots of insight into just how polynomial equations work, and it proves the point, if somewhat laboriously. The second uses a more modern approach, typifying contemporary algebra in drawing the conclusion as a by-product of much more inclusive theory.

Exercise 7–20.   Construct a polynomial in $Q[x]$ having $2 + 3i$ and $\sqrt{2}$ as zeros. Does it have to have other zeros? How many zeros must it have? How many may it have?

Exercise 7–21.   Construct a polynomial having θ_1, θ_2, and θ_3 as zeros. Show that every transposition (interchange) of two of the zeros leaves the coefficients of the polynomial fixed.

CHAPTER 8

GALOIS THEORY

"Beautiful," "structural," "rich," "elegant"—these are words used in various introductions to Galois theory, where the theory of groups and the theory of fields are linked so as to explain each other. The theory is quite capable of selling itself, so let us introduce it first and then admire it.

Definition 8–1. An **automorphism** σ of a mathematical system S with an operation is a one-to-one correspondence from S onto itself that is an isomorphism with respect to the operation.

Definition 8–2. An **automorphism** σ of a field F is an automorphism of F that preserves both field operations. An element $f \in F$ is **fixed** by an automorphism σ if σ maps it to itself: $\sigma f = f$; it is **moved** by σ if $\sigma f \neq f$.

Exercise 8–1. Show that the correspondence $i \leftrightarrow -i$ provides an automorphism of the complex number field C. What is the image of $s + ti$? Show

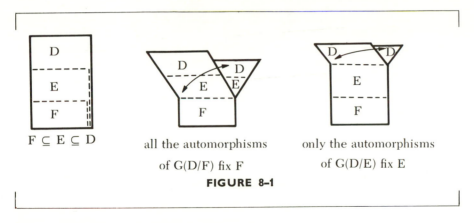

$F \subseteq E \subseteq D$

all the automorphisms
of G(D/F) fix F

only the automorphisms
of G(D/E) fix E

FIGURE 8–1

that corresponding complex numbers are conjugates (Definition 7–5). Show that each element of R is fixed by the automorphism.

Theorem 8–1. Let D be a finite extension of F, and let E be an intermediate field $F \subseteq E \subseteq D$. Let $G(D/F)$ be the automorphisms of D that fix every $f \in F$. Let $G(D/E)$ be the automorphisms of D that fix every $e \in E$. Then $G(D/F)$ is a group under composition and $G(D/E)$ is a subgroup of $G(D/F)$. Let H be a subgroup of $G(D/F)$. Then the elements of D that are fixed by every $h \in H$ form an intermediate field E^H with $F \subseteq E^H \subseteq D$ (shown schematically in Figure 8–1).

PROOF: First we prove that $G(D/F)$ is a group under composition. If σ_1 and σ_2 are two automorphisms in $G(D/F)$ then any element $d \in D$ has an image $\sigma_2 d$ in D, which in turn has an image $\sigma_1(\sigma_2 d)$ in D. Then the composition of the two, $\sigma_1\sigma_2$, produces an element that lies in D. Both σ_1 and σ_2 preserve the two field operations since they are field automorphisms, so $\sigma_1\sigma_2(d_1 * d_2) = \sigma_1(\sigma_2 d_1 * \sigma_2 d_2) = \sigma_1(\sigma_2 d_1) * \sigma_1(\sigma_2 d_2) = \sigma_1\sigma_2 d_1 * \sigma_1\sigma_2 d_2$ for each field operation $*$. Any $f \in F$ is fixed by σ_2, then fixed by σ_1, hence fixed by $\sigma_1\sigma_2$. Then $G(D/F)$ is closed under the binary operation composition. The identity correspondence provides the identity $\hat{1}$ of $G(D/F)$, for $\hat{1}(\sigma d) = \sigma d$ and $\sigma(\hat{1}d) = \sigma d$. If an automorphism σ maps $d \in D$ to σd then the mapping $\sigma d \to d$ for all σd in D is also an automorphism of D that fixes every $f \in F$, and so is the inverse automorphism σ^{-1}. Then $G(D/F)$ is a group and similarly $G(D/E)$ is a group under the same operation of composition.

Each element of $G(D/E)$ is in $G(D/F)$, for if σ fixes all elements of E it certainly fixes all elements of $F \subseteq E$. Then $G(D/E)$ is a subgroup of $G(D/F)$.

Now suppose H is a subgroup of $G(D/F)$. Each automorphism σ of H is in $G(D/F)$, so that it fixes all elements f of F. Then the elements of D fixed by all automorphisms $\sigma \in H$ include all $f \in F$. Do they form a field? They include 0 and 1, which are elements of F. Suppose e_1 and e_2 are fixed by σ. Then if $*$ is one of the field operations, $\sigma(e_1 * e_2) = (\sigma e_1) * (\sigma e_2) = e_1 * e_2$ so that $e_1 * e_2$ is also fixed by σ; that is, the elements of D that are fixed by all automorphisms $\sigma \in H$ are closed under the binary operations of the field. If e is fixed by σ, then from $e + (-e) = 0$ we have $\sigma[e + (-e)] = \sigma e + \sigma(-e) = e + \sigma(-e) = \sigma(0) = 0$ so that $\sigma(-e) = -e$. Similarly, if $e \neq 0$, $e \cdot e^{-1} = 1$, so that $\sigma[e \cdot e^{-1}] = (\sigma e)(\sigma e^{-1}) = e(\sigma e^{-1}) = \sigma(1) = 1$, or $\sigma e^{-1} = e^{-1}$. Each operation is commutative, since it is commutative in D, and the associative and distributive laws are also inherited from D. ∎

Definition 8–3. Let E be the splitting field (unique up to an isomorphism by Theorem I–2) of a polynomial $f(x) \in F[x]$. The group $G(E/F)$ of automorphisms of E that fix every $f \in F$ is the **Galois group** or **group of the polynomial** and the **group of the equation** $f(x) = 0$.

Exercise 8–2. Let F be the rationals Q, and let $f(x)$ be the quadratic polynomial $ax^2 + bx + c$, $a \neq 0$. Let θ and $\bar{\theta}$ be the zeros of $f(x)$ in the splitting field E. We can show (Exercises 7–17 and 7–18) that $\bar{\theta} \in Q(\theta)$, so $E \cong Q(\theta)$. With 1 and θ as basis elements, every $e \in Q(\theta)$ can be expressed as $s + t\theta$, where $s, t \in Q$. From this show that every automorphism σ of $G(Q(\theta)/Q)$ is determined completely by $\sigma\theta$. Deduce that $G(Q(\theta)/Q)$ has two elements, $\hat{1}$ and $\sigma: \theta \leftrightarrow \bar{\theta}$, in case $\theta \notin Q$, otherwise only the identity element $\hat{1}$.

Theorem 8–2. Let F be a field. Let $[F(\theta):F] = n$. Any automorphism σ in $G(F(\theta)/F)$ is completely determined by $\sigma(\theta)$. The order of $G(F(\theta)/F)$ is less than or equal to n.

Let E be the splitting field of an irreducible polynomial $p(x) \in F[x]$. The Galois group $G(E/F)$ is isomorphic to a group of permutations of the k zeros of $p(x)$ in E, hence to a subgroup of the symmetric group S_k.

This theorem links the study of groups to the permutations of the roots of an equation. Historically, the development of groups in the abstract sense of Chapter 1 came long after Abel and Galois studied the particular group formed by permuting the roots of an equation. It was Kronecker who, in 1870, isolated the group properties in their own right.

PROOF: By Theorem 7–4 θ is algebraic relative to F. Let $p(x)$ be the monic irreducible polynomial in $F[x]$ having θ as a zero (Theorem 7–1).

Suppose deg $(p(x)) = k$. Then, from Theorem 7–4, $1, \theta, \theta^2, \ldots \theta^{k-1}$ form a linear basis for $F(\theta)$ over F, so $k = n$, and each element ψ in $F(\theta)$ can be written

$$\psi = \sum_{i=0}^{n-1} f_i \theta^i, \quad f_i \in F.$$

Since σ is an automorphism, it preserves addition and multiplication. Since σ fixes F, we have $\sigma f_i = f_i$ for each coefficient. Then

$$\sigma(\psi) = \sigma\left(\sum_{i=0}^{n-1} f_i \theta^i\right) = \sum_{i=0}^{n-1} (\sigma f_i)(\sigma \theta^i) = \sum_{i=0}^{n-1} f_i (\sigma \theta)^i$$

is completely determined by the image $\sigma \theta$ of θ.

Applying σ to the equal elements

$$p(\theta) = 0,$$

we have

$$\sigma(p(\theta)) = \sigma(0) = 0,$$

but since σ preserves $+$ and $\cdot$,

$$\sigma(p(\theta)) = p(\sigma(\theta)),$$

so that

$$p(\sigma(\theta)) = 0$$

for each $\sigma \in G(F(\theta)/F)$. Each $\sigma(\theta)$ is one of the zeros of $p(x)$, of which there are at most n, by Lemma 7–2. Then $|G(F(\theta)/F| \le n$.

Now let E be the splitting field of an irreducible polynomial $p(x)$ in $F[x]$. From Theorem 7–5 we know that the splitting field $E \cong F(\theta_1, \theta_2, \ldots, \theta_k)$, where the θ's are the zeros of $p(x)$ in E. If $\sigma \in G(E/F)$, then from $p(\theta_i) = 0$ we have

$$\sigma[p(\theta_i)] = p(\sigma \theta_i) = \sigma(0) = 0,$$

so that $\sigma \theta_i$ is a conjugate zero θ_j of $p(x)$. By Theorem 7–3ii each element e of E is a rational combination of the θ's, so the effect on e of any automorphism σ in $G(E/F)$ is determined by which permutation σ effects on the θ's. From Theorem 8–1 the automorphisms in $G(E/F)$ form a group; hence so do the permutations to which they correspond. Hence the latter form a subgroup of the symmetric group S_k of permutations on k letters. ∎

Exercise 8–3. Let $F = Q$, $D = Q(\sqrt{2}, i)$, $E_1 = Q(\sqrt{2})$, and $E_2 = Q(i)$. Then F has characteristic 0, and $D = F(\psi)$, where $\psi = \sqrt{2} + fi$, $f \ne 0 \in Q$ (see Figure I–2). Find $\sigma_1 \psi$ for each automorphism $\sigma_1 \in G(D/E_1)$ and $\sigma_2 \psi$ for each $\sigma_2 \in G(D/E_2)$. Find the four conjugates of $\psi = \psi_1$, hence, by Theorem 8–2, the four automorphisms of $G(D/F)$.

Definition 8–4.   A finite extension E of F is a **normal extension** if every irreducible polynomial $p(x) \in F[x]$ with one zero in E has all its zeros in E (splits in E).

Theorem 8–3.   If E is a normal extension of F, a field of characteristic zero, then $|G(E/F)| = [E:F]$.

PROOF: Since E is normal over F, it is a finite extension. Let $[E:F] = n$. Then from the Corollary to Theorem I–4 E has a basis $1, \psi, \psi^2, \ldots, \psi^{n-1}$ over F, where ψ is a zero of an irreducible polynomial of degree n in $F[x]$. Since E is normal, all the conjugates of ψ lie in E. They are all distinct, from Lemma 7–2. Then each correspondence $\psi \leftrightarrow \psi_j$ between ψ and one of its conjugates (including itself) provides an automorphism of $G(E/F)$, and there are n of these. Then $|G(E/F)| \geq n$, and from Theorem 8–2, $|G(E/F)| = n$. ∎

Definition 8–5.   Let H be a group of automorphisms of a field D. The subfield $E \subseteq D$ of all elements that are fixed by every automorphism in H is called the **fixed field** of H.

At this point we postpone complete proofs until Appendix J, simply outlining here the development of the Fundamental Theorem of Galois Theory.

When the base field F contains the rationals Q so that F is a "field of characteristic zero," the definition (Definition 8–4) adopted here for a *normal* extension field is just one of three necessary and sufficient conditions that could be taken as definitive.

Whenever we have "necessary-and-sufficient" or "if-and-only-if" conditions, they characterize the object completely and so can serve to define it mathematically.

Combining Definition 8–4 with Theorems J–1 and J–2, and letting F be a field of characteristic zero and E be a finite extension of F, we have the three

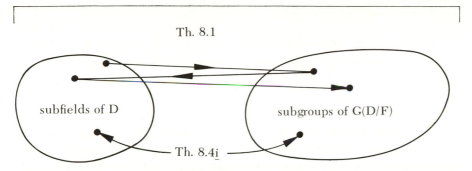

Th. 8.1

subfields of D   subgroups of G(D/F)

Th. 8.4i

FIGURE 8–2

By Theorem 8–1 each subfield defines a subgroup and each subgroup defines a subfield. In general the correspondence is not one-to-one. By Theorem 8–4i if D is normal over F, then the correspondence is one-to-one.

conditions:

E is a normal extension $\longleftrightarrow$

1. every irreducible polynomial with one zero in E splits in E
2. F is the fixed field of $G(E/F)$
3. E is the splitting field over F for some polynomial in $F[x]$.

Condition 2 provides a clue to the way the two mappings of Theorem 8–1 are tightened for normal extensions of fields of characteristic zero into the single one-to-one correspondence exploited in Theorem 8–4: How could F *fail* to be the fixed field of $G(E/F)$ when Theorem 8–1 defines $G(E/F)$ as the automorphisms that fix F? Unless E is normal over F, the automorphisms that fix F also fix other elements of E that are not in F. If D is an extension of F, then Theorem 8–1 guarantees for each intermediate field E, with $F \subseteq E \subseteq D$, a subgroup $G(D/E)$ of $G(D/F)$ and for each subgroup H of $G(D/F)$ some intermediate field F^H. If we take D to be a normal extension, the fixed field is completely determined, enabling us to deduce the Fundamental Theorem of Galois Theory.

Theorem 8–4. (Fundamental Theorem of Galois Theory). Let D be a normal extension of F, a field of characteristic zero. Let E be an intermediate field $F \subseteq E \subseteq D$. Then the following four statements hold:

i. $E \longleftrightarrow G(D/E)$ provides a one-to-one correspondence between all the intermediate fields E and all the subgroups of $G(D/F)$.

ii. $[D:E] = |G(D/E)|$ and $[E:F] = [G(D/F):G(D/E)]$.

iii. $D \supseteq E_2 \supset E_1 \supseteq F$ if and only if $\hat{1} \subseteq G(D/E_2) \subset G(D/E_1) \subseteq G(D/F)$. (Note $E_2 \supset E_1$ indicates "proper" inclusion; E_2 has elements that are not elements of E_1.)

iv. E is a normal extension of F if and only if $G(D/E)$ is a normal subgroup of $G(D/F)$, in which case $G(E/F) \cong G(D/F)/G(D/E)$.

PROOF: See Appendix J.

Classroom Exercise 8–4. Continuing Exercise 8–3, we have $F = Q$, $D = Q(\sqrt{2}, i)$. In this case D is a normal extension of F, so we can apply Theorem 8–4. Take $\psi = \sqrt{2} + fi$, $f \neq 0 \in Q$. As shown in Theorem I–4, $D = Q(\sqrt{2}, i) = Q(\psi)$. Let the automorphisms of $G(D/F)$ be $\hat{1} = (\psi \leftrightarrow \psi_1 = \psi)$, $a = (\psi \leftrightarrow \psi_2 = \sigma_1\psi)$, $b = (\psi \leftrightarrow \psi_3 = \sigma_2\psi)$, and $c = (\psi \leftrightarrow \psi_4 = \sigma_1\sigma_2\psi)$. Then the Cayley composition table for $G(D/F)$ is the same as that for the four-group.

Find the unique subgroup of $G(D/F)$ that corresponds to each intermediate field $E_1 = Q(\sqrt{2})$ and $E_2 = Q(i)$. Also, find the subfield corresponding to each subgroup.

∘	1	a	b	c
1	1	a	b	c
a	a	1	c	b
b	b	c	1	a
c	c	b	a	1

Classroom Exercise 8–5. Continue the previous exercise, verifying that part *ii* of Theorem 8–4 holds in this case.

Classroom Exercise 8–6. Continuing the previous exercise, show that $\hat{1} \subset G(D/E) \subset G(D/F)$ for $E = E_1$ and for $E = E_2$, but that $G(D/E_2) \not\subset G(D/E_1)$. Why are the inclusions reversed in order in the second clause of Theorem 8–4*iii*?

Classroom Exercise 8–7. Continuing Exercises 8–3 to 8–6, show that $G(D/E_1)$ is normal in $G(D/F)$ and that $G(E_1/F) \cong G(D/F)/G(D/E_1)$, and similarly for E_2.

The following sequence of exercises concerns the polynomial function $f(x) = x^3 + 3x^2 + 4$. (Also see Exercise I–4.) The function has a real zero $\theta_1 = a + b - 1$, where $a = (-3 + \sqrt{8})^{1/3}$ and $b = (-3 - \sqrt{8})^{1/3}$, and two complex zeros, θ_2 and θ_3.

Exercise 8–8. Prove that $f(x)$ is irreducible in $Q[x]$. Thus demonstrate that there is an irreducible polynomial with a zero in $E = Q(\theta_1)$ that fails to split in E. From this deduce that E is not a normal extension of Q.

Exercise 8–9. Use Theorem 7–5 to establish that $D = Q(\theta_1, \theta_2) = Q(\theta_1, \theta_2, \theta_3)$ is a splitting field for $f(x)$ over Q. Then use one of the three necessary and sufficient conditions to prove that D is a normal extension of Q.

Exercise 8–10. Use Theorem 7–3*iii* to find $[E:Q]$. From Exercises 8–8 and 8–9 conclude that $D \supset E$. From Theorem 8–2 show that $[D:Q] \mid 3! = 6$. Find $[D:Q]$.

Exercise 8–11. Use Exercise 8–10 and Theorem 8–2 to argue that the automorphisms of $G(D/F)$ are $\hat{1}$, (23), (12), (13), (123), and (132), where (13) stands for the automorphism of D that makes θ_1 correspond to θ_3, (132) stands for the correspondence

$$\theta_1 \leftrightarrow \theta_3$$

$$\theta_3 \leftrightarrow \theta_2$$

$$\theta_2 \leftrightarrow \theta_1,$$

and so on. The automorphisms of $G(D/E)$ are $\hat{1}$ and (23). From Theorem 8–4*ii* find $[D:E]$. Show that $G(D/E)$ is not normal in $G(D/Q)$, as expected from Theorem 8–4*iv*.

CHAPTER 9

RADICALS AND ROOTS OF UNITY

Abel's proof of the theorem named for him was less than perfect, because of the way the roots of unity were handled. Here we study these roots and in the next chapter adjoin the ones we need to the coefficient field of the equation to be solved.

> **Definition 9–1.** Let F be a field of characteristic zero, and let D be an extension of F in which $x^n - 1$ splits. The n zeros of $x^n - 1$ in D are the **nth roots of unity.** (By Lemma 7–2 there can be only n roots of unity.)

When we solve an equation "by radicals," we extend the coefficient field F sequentially by adjoining zeros of polynomials like $x^n - f$. We are led to study nth roots of unity r, because if w is a zero of $x^n - f$, then so is wr^i, for i a positive integer, because

$$(wr^i)^n - f = w^n(r^i)^n - f = fr^{in} - f = f(r^n)^i - f = f(1)^i - f = 0.$$

Theorem 9–1. The nth roots of unity are distinct and form a cyclic group under multiplication.

PROOF: This result from number theory is proved in Appendix K.

Definition 9–2. A generator of the cyclic group of the nth roots of unity is a **primitive nth root of unity.**

Exercise 9–1. Find the two square roots of unity. Show that they form a cyclic group.

Exercise 9–2. Draw in the complex plane, as in Figure 9–1, the 3rd roots of unity, $e^{2k\pi i/3}$, $k = 0, 1, 2$. Complete an equilateral triangle of unit side with an altitude a, as in Figure 9–2. Use the Theorem of Pythagoras to find a; then find the 3rd roots of unity in the form $s + ti$. Compare with the 6th roots of unity shown in Figure 9–1.

Theorem 9–2. Let F be a field of characteristic 0. Let $E_0 = F(r_1, r_2, \ldots, r_n)$, where $r_1, r_2, \ldots, r_n$ are all the nth roots of unity. Then $E_0 = F(r)$, where r is a primitive nth root of unity. Also, E_0 is a normal extension of F, and $G(E_0/F)$ is abelian.

PROOF: From Theorem 9–1 the roots $r_1, r_2, \ldots, r_n$ form a cyclic group with a generator r, a primitive nth root of unity. Then $r \in F(r_1, r_2, \ldots, r_n)$. Since r generates all the roots, $r_i \in F(r)$, $i = 1, 2, \ldots, n$. Then $E_0 = F(r)$. E_0 is normal over F, as it is the splitting field of $x^n - 1$ (Theorems 7–5 and J–2).

From Theorem 7–4 r is algebraic relative to F. Then by Theorem 7–1, r is a zero of a unique monic irreducible polynomial of minimal degree, $\Phi_n(x) \in F[x]$, which divides each polynomial having r as a zero. Then $\Phi_n(x) \mid (x^n - 1)$.

From Theorem 8–2 the automorphisms σ of $G(E_0/F)$ are determined by the conjugates σr of r in $\Phi_n(x)$. Because $\Phi_n(x) \mid (x^n - 1)$, each conjugate is an nth root of unity. If $[\sigma r]^e = \sigma(r^e) = \sigma(1) = 1$, then $r^e = 1$, since σ is an automorphism, so $e = n$, and $\sigma(r)$ is primitive. Since r generates the nth roots of unity, each primitive nth root can be written as a power of r, r^y. Then $(y, n) = 1$, for if $(y, n) = d > 1$, then $(r^y)^{n/d} = (r^n)^{y/d} = 1$, so that r^y would not be primitive.

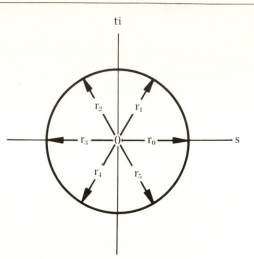

The six distinct complex 6th roots of unity are

$$e^{k2\pi i/6}, \quad k = 0, 1, 2, 3, 4, 5,$$

for $(e^{k2\pi i/6})^6 = e^{k2\pi i} = \cos k2\pi + i \sin k2\pi = 1$. From $e^{i\theta} = \cos\theta + i\sin\theta$, we can also write the roots

$$r_1 = \frac{1 + \sqrt{3}\,i}{2}, \qquad r_4 = \frac{-1 - \sqrt{3}\,i}{2},$$

$$r_2 = \frac{-1 + \sqrt{3}\,i}{2}, \qquad r_5 = \frac{1 - \sqrt{3}\,i}{2},$$

$$r_3 = -1, \qquad r_6 = 1.$$

r_1 and r_5 are primitive.

In general the complex nth roots of unity are $e^{k2\pi i/n}$, $k = 0, 1, \ldots, n - 1$. The primitive roots are those for which $(k, n) = 1$.

FIGURE 9–1

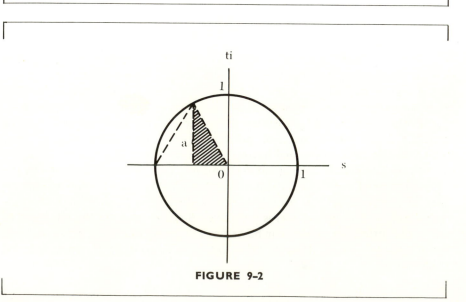

FIGURE 9–2

Let σ and τ be two members of $G(E_0/F)$. Then the effect of σ on r can be given as

$$\sigma r = r^y,$$

for some y with $(y, n) = 1$, and also

$$\tau r = r^z,$$

for some z with $(z, n) = 1$. We have

$$(\sigma\tau)r = \sigma(\tau r) = \sigma(r^z) = (r^y)^z = r^{yz},$$

and

$$(\tau\sigma)r = \tau(\sigma r) = \tau(r^y) = (r^z)^y = r^{zy} = r^{yz}$$

$((yz, n) = 1)$. Since by Theorem 7–4 E_0 has a basis in terms of powers of r and 1, the fact that $(\sigma\tau)r = (\tau\sigma)r$ implies that $\sigma\tau = \tau\sigma$ on all of E_0. Then $G(E_0/F)$ is abelian. ∎

Corollary. Let $E_0 = F(u_0, u_1, \ldots, u_{k-1})$, where u_i is a primitive n_ith root of unity. Let n be the least common multiple of the n_i, $n = [n_0, n_1, \ldots, n_{k-1}]$, and let u be a primitive nth root of unity. Then $E_0 = F(u)$ is normal over F and $G(E_0/F)$ is abelian.

PROOF: Since $n = [n_0, n_1, \ldots, n_{k-1}]$, n is a multiple $m_i n_i$ of each n_i. Then u^{m_i} has order n_i and, hence, generates the n_ith roots of unity. Thus $u_i \in F(u)$ for each $i = 0, 1, \ldots, k - 1$. Since the product $u_0 u_1 \cdots u_{k-1}$ has order n, it generates the nth roots of unity, including u, so $u \in E_0$. Thus $E_0 = F(u)$ and the rest of the Corollary follows from Theorem 9–2. ∎

Theorem 9–3. Let E_0 be a field of characteristic zero containing all the nth roots of unity. Let $f \in E_0$, and let E be a splitting field for $x^n - f$ over E_0. If w is any zero of $x^n - f$ in E, then $E = E_0(w)$, and $G(E/E_0)$ is cyclic.

PROOF: Let w be one zero of $x^n - f$ in E, and let r be a primitive nth root of unity. Then the n distinct zeros of $x^n - f$ are $w, wr, wr^2, \ldots, wr^{n-1}$, which all lie in $E_0(w)$. Since w lies in any subfield of E in which $x^n - f$ splits, $E = E_0(w)$.

Each automorphism $\sigma_i \in G(E/E_0)$ sets up a correspondence

$$\sigma_i(w) = wr^{k_i},$$

for some k_i with $0 \leq k_i < n$. This correspondence preserves composition and multiplication, respectively, for

$$\sigma_i \sigma_j(w) = \sigma_i(\sigma_j w) = \sigma_i(wr^{k_j}) = (wr^{k_j})r^{k_i} = wr^{k_i+k_j}.$$

(Here we use the fact that each r^k is an element of E_0, the fixed field of $G(E/E_0)$.) Then $G(E/E_0)$ under composition is isomorphic to a subgroup of the cyclic group of the nth roots of unity. Since a subgroup of a cyclic group is cyclic, $G(E/E_0)$ is cyclic. ∎

Exercise 9–3. Explain why a subgroup of a cyclic group is cyclic.

Exercise 9–4. Explain why an isomorphic copy of a cyclic group is cylic.

Theorem 9–4.   Let $f \neq 0$ be an element of F, and let $x^n - f$ split in D. Then D contains the nth roots of unity.

PROOF: Let w be one of the zeros of $x^n - f$ in D. Then from Theorem 9–3, $F(w, wr, wr^2, \ldots, wr^{n-1})$ is a splitting field for $x^n - f$ over F (Theorem 7–5). This field may fail to be a subfield of D. However, D contains a splitting field for $x^n - f$ isomorphic to $F(w, wr, wr^2, \ldots, wr^{n-1})$ in which $x^n - f$ has zeros $w, z_1, z_2, \ldots, z_{n-1}$ (Theorem I–2). If z_1 corresponds to wr under such an isomorphism, then z_1/w corresponds to r and, hence, to a primitive nth root of unity and, therefore, is a primitive nth root of unity. Then D contains all the nth roots of unity. ∎

Exercise 9–5. In Theorem 9–3 let $E_0 = Q$, $n = 2$, $f = 7$. Find the splitting field E for $x^2 - 7$ over Q. Show that $G(E/E_0)$ has order 2.

Exercise 9–6. In Theorem 9–3 let $E_0 = Q$, $n = 2$, and $f = 4$. Show that $x^2 - f$ in this case splits in E_0, so that $G(E/E_0)$ has only the identity automorphism.

Exercise 9–7. In Theorem 9–3 let $E_0 = Q(r)$, where r is a primitive 3rd root of unity, $n = 3$, $f = 7$. Factor $x^3 - f$ into linear factors in its splitting field over E_0. Show that the Galois group $G(E/E_0)$ is cyclic of order 3.

Exercise 9–8. Show that the Galois group for $x^3 - 7$ over Q, $G(E/Q)$, is isomorphic to S_3 and, hence, is not abelian (cf. Exercise 9–7).

Exercise 9–9. In Theorem 9–3 let $E_0 = Q(r)$, where r is a primitive 3rd root of unity, $n = 3$, and $f = 8$. Show that $G(E/E_0)$ has order 1.

CHAPTER 10

SOLUTION BY RADICALS

By completing the square we derived formula solutions for the two roots of a quadratic equation $x^2 + bx + c$:

$$\theta_1 = \frac{-b + \sqrt{b^2 - 4c}}{2} \quad \text{and} \quad \theta_2 = \frac{-b - \sqrt{b^2 - 4c}}{2}.$$

The calculation of the roots involves finding a zero w of $x^2 - f$, where $f = b^2 - 4c$, and performing rational operations on w and the coefficients b, c.

Given a cubic equation we can make it monic by multiplying throughout by the inverse of the leading coefficient. Then if the coefficient of the 2nd degree term is b, we can make a change of variable from y to $x - b/3$, which will eliminate the second degree term. Thus we can transform the general cubic to the form

$$x^3 + px + q = 0.$$

Exercise 10–1. Perform this change on the cubic equation

$$y^3 + 3y^2 + 4 = 0.$$

Formulas, variously attributed, can be used to give the 3 roots of the cubic: Let r stand for a primitive cube root of unity. Then the roots of $x^3 + px + q$ are

$$\theta_1 = \frac{A^{1/3} + B^{1/3}}{3}, \quad \theta_2 = rA^{1/3} + r^2B^{1/3}, \quad \theta_3 = r^2A^{1/3} + rB^{1/3},$$

155

where $A = (-27q + 3\sqrt{-3}\,\Delta)/2$, $B = (-27q - 3\sqrt{-3}\,\Delta)/2$, and $\Delta = \sqrt{-(27q^2 + 4p^3)}$. The choice of Δ between the two possible square roots and the choice among the three possible cube roots of A and B must be made so that $(AB)^{1/3} = -3p$.

The calculation of the roots of the cubic involves finding a zero w_1 of $x^2 - f_0$, where $f_0 = -(27q^2 + 4p^3)$, finding a zero w_2 of $x^2 - f_2$, where $f_2 = -3$, performing rational operations on w_1, w_2, and the coefficients p, q to obtain A and B, then finding a zero w_3 of $x^3 - A$ and a zero w_4 of $x^3 - B$. (Theorem 9–4 shows that the cube roots of unity can be expressed in terms of w_1, w_2, w_3, and w_4.)

Exercise 10–2. Use the preceding formulas to find the real root θ_1 of $x^3 - 3x + 6 = 0$. Compute $y_1 = \theta_1 - 1$ correct to one decimal place and see how well it satisfies the equation of Exercise 10–1.

The general fourth degree equation can also be solved by formula (see J. V. Uspensky, *Theory of Equations*, McGraw-Hill, 1948, pp. 94–98) but not the general equation of fifth degree. Many individual quintics can be solved by rational operations and root extractions starting with their coefficients, but no general formula can be given for the roots in terms of the coefficients, like those available for equations of degree < 5.

> **Definition 10–1.** Let F be a field of characteristic zero. Let $f(x)$ be a polynomial of positive degree in $F[x]$. The polynomial equation $f(x) = 0$ is **solvable by radicals** if all its roots can be calculated from its coefficients in a finite number of steps by rational operations and root extractions. **Rational operations** are the field operations $+$ and $\cdot$. An **nth root** w of a field element f is an element in an extension field $E \supseteq F$ for which $w^n = f$.

We immediately rephrase this idea in terms of field extensions, defining a root tower.

> **Definition 10–2.** Let F be a field of characteristic zero, let $f(x)$ be of positive degree in $F[x]$, and let $f(x) = 0$ be **solvable by radicals.** Then all the roots of $f(x) = 0$ lie in a field F_s, where
>
> $$F = F_0 \subseteq F_1 \subseteq \cdots \subseteq F_s,$$
>
> $F_{i+1} = F_i(w_{i+1})$, and $w_{i+1}^{n_i} = f_i \in F_i$, $i = 0, 1, \ldots, s - 1$. The sequence of field extensions is a **root tower** for F_s over F.

Exercise 10–3. Write out in order the equations $x^{n_i} - f_i$ with w_{i+1} as a zero for the case of the general cubic. Notice that $f(y) = (y - 1)(y^2 - 2)$ splits in R, but the translated $f(x + \frac{1}{3})$ yields $A = -17 - 3\sqrt{6}\, i \notin R$.

Definition 10–3. Let G be a finite group. G is **solvable** if there is a chain of subgroups

$$G = G_0 \rhd G_1 \rhd \cdots \rhd G_t = \langle 1 \rangle,$$

such that G_i/G_{i+1} is abelian for $i = 0, 1, \ldots, t - 1$, where the **symbol** "$\rhd$" means "has the normal subgroup."

In Appendix L we prove, by three theorems and a lemma, that if a polynomial is solvable by radicals, then its Galois group is solvable. You can see in a general way from Chapters 8 and 9 why this is so: Suppose the equation is known to be solvable by radicals. This means, according to Definition 10–2, that a tower of extensions can be built by adjoining n_ith roots until the polynomial splits in some field F_s. The sequence of extension fields suggests the existence of a corresponding sequence of nested subgroups of $G(F_s/F)$, such as those of Definition 10–3, but we lack a basic hypothesis of the Fundamental Theorem 8–4 of Galois Theory that the extension be normal. Also, it would be convenient to have all the relevant roots of unity in the base field F, because then Theorem 9–3 would imply that each quotient group in Definition 10–3 was cyclic, which in turn would imply that it was abelian.

We get around these difficulties in Appendix L by replacing the tower of fields in Definition 10–2 by a tower of normal field extensions, then replacing this second tower by a third built not on F itself but on F with all the relevant roots of unity adjoined. Then we can use Theorems 8–4 and 9–3, but on the wrong fields! Eventually, we show that the Galois group is a homomorphic image of a group which we can prove to be solvable and that this fact implies that the Galois group is also solvable.

We do not prove it here, but the converse also holds, so there is a necessary and sufficient condition: "A polynomial equation is solvable by radicals if and only if its Galois group is solvable."

Now you have enough background to appreciate our goal theorem and help to prove it.

Theorem 10–1 (Abel's Theorem). The general polynomial equation of degree >4 is not solvable by radicals.

PROOF: We prove there can be no general formula solution by exhibiting a quintic equation $f(x) = 0$ whose Galois group is not solvable. Then by Theorem L–3, $f(x) = 0$ is not solvable by radicals.

Exercise 10–4.   Prove that there can be no general formula for solving equations of degree >5 if there can be none for degree $= 5$.

Let $f(x) = x^5 - 4x^4 + 2x + 2 \in Q[x]$. We have already built up quite a lot of information about this polynomial. In Exercise 6–22 we showed that it had 3 real zeros, in Exercise 6–31 that it had 2 complex zeros, and in Exercise 6–28 that it was irreducible.

From Theorem 8–2 we know that the Galois group G of $f(x)$ is isomorphic to a subgroup of S_5, the permutations of the 5 zeros of $f(x)$. We shall prove that $G \cong S_5$.

First we establish that **G contains a 5-cycle**. From Theorem 8–3, $|G| = [D:Q]$, where $D = Q(\theta_1, \theta_2, \theta_3, \theta_4, \theta_5)$, the θ's representing the 5 zeros of $f(x)$. From Theorems 7–5 and I–3, $[D:Q] = [Q(\theta_1):Q][D:Q(\theta_1)] = 5[D:Q(\theta_1)]$. Then the order of G has a divisor 5. Then from Theorem 5–3, G must have a subgroup of order 5. Since by Theorem 5–2 this subgroup must be cyclic, we know that G must contain a permutation of the roots θ_i whose order is 5. This permutation must be a 5-cycle, since any product of disjoint cycles of the 5 roots fails to have order 5.

Next we establish that **G contains a transposition**. One permutation known to be an element of G is the transposition $(\theta_1\theta_2)$, where we choose the first two roots to be the two complex ones: For suppose $\theta_1 = s + ti$. Then θ_1 is a zero of a monic irreducible polynomial in $R[x]$ of minimal degree. In fact, θ_1 is a zero of $[x - (s + ti)][x - (s - ti)] = x^2 - 2sx + s^2 + t^2$ and of no polynomial of first degree since $\theta_1 \notin Q$. Then this polynomial and its factor $x - (s - ti)$ divide $f(x)$ in $D[x]$, so that $s - ti$, the complex conjugate of θ_1, is the other complex zero of $f(x)$. This transposition $(\theta_1\theta_2)$ in D replaces each element d by its complex conjugate, which also lies in D as the other zero of the irreducible polynomial having d as zero.

Now it is a matter of computation to show that since G contains a 5-cycle of the roots of $f(x) = 0$ and a transposition, G must contain every permutation of the roots. We need only use the facts that G is closed under its operation, composition, and contains all inverses. Dropping the θ's and using only the numerical subscripts, let the transposition be (12) and the 5-cycle be $(abcde)$, where 1 and 2 are necessarily equal to two of these letters. We can arbitrarily start the cycle with whatever letter equals 1, following with the other letters in order. If 2 then occupies the kth place, $k = 2, 3, 4$, or 5, then $(abcde)^{k-1}$ has the form $(12---)$.

Exercise 10–5.   Verify the last sentence.

Exercise 10–6.   Let $u = (12)$ and $v = (12345)$. Find vuv^{-1}, v^2uv^{-2}, and v^3uv^{-3}.

Exercise 10–7. Supposing now that G contains u, v, and (23), (34), and (45), find the following, which by closure it also contains:

$$
\begin{array}{ll}
(23)(12)(23) & (34)(23)(34) \\
(34)(13)(34) & (45)(24)(45) \\
(45)(14)(45) & (45)(34)(45)
\end{array}
$$

From Exercises 10–6 and 10–7 we show that G contains all the transpositions of the 5 symbols taken 2 at a time. Then by Theorem 3–3, **G contains all of S_5.**

However, **S_5 is not solvable.** The only chain of subgroups, each normal in the preceding, for S_5 is $S_5 \rhd A_5 \rhd \langle 1 \rangle$, since by Theorem 5–4, A_5 is simple.

Exercise 10–8. Show that if S_5 has a normal subgroup S, then $A_5 \cap S$ is a normal subgroup of A_5. (Borrow from the proof of Lemma L–1.)

Since $A_5/\langle 1 \rangle \cong A_5$ is not abelian, S_5 is not solvable. By Theorem L–3, $f(x) = 0$ is not solvable by radicals. ∎

Exercise 10–9. Prove that A_5 is not abelian. Why does the chain $S_5 \rhd \langle 1 \rangle$ not provide a solvability chain for S_5?

We are now in a position to appreciate the Galois theory, having explored its main points and then applied them to a problem (solution of equations). When we actually solve equations we may use a wide variety of methods, formulas, graphs, numerical approximations, and so forth. This often leads the student of high school algebra to an impression of algebra as a heterogeneous collection of tricks. However, as we see from this theory, much of algebra can be reasoned closely from a set of definitions and postulates, just as plane geometry can be. For many (certainly not all), it is esthetically pleasing to discover that what seemed to be unrelated techniques in elementary algebra really proceed from a few underlying principles. It provides an answer at last to the question that nags us in elementary algebra, "WHY?"

APPENDICES

Sometimes we try to cover so much in a mathematics course that we discover only the quick facile student, overlooking the talent of a slower but sound student or one who panics at class recitations or tests. The Appendices supply a welcome shift of emphasis for such students, who may prepare reports on them in written or in oral form or construct diagrams or models. Here instructor and students have a chance to discover latent talent for the deductive logic of proof or for mathematical bibliography. Some of the topics may suggest projects for teachers to supervise for Science Fairs or Exhibits later.

LIST OF APPENDICES, pertinent part of text

Q HAS $\aleph_0$ MEMBERS

How many rational numbers are there? We know that Q has at least as many members as I, for the correspondence used to show $Q_1 \cong I$ also serves to count Q_1. If we let q stand for the number of rational numbers in Q, then we have $q \geq \aleph_0$.

There is an ingenious way to show that $q \leq \aleph_0$. We set up a correspondence between integers and number pairs. Each member of Q, a/b, is represented ($\aleph_0$ times) as the solution of an equation $bx = a$, so we count these equations. In this way we count all the rational numbers, not once, but many times, since a/b occurs as the solution of every equation $bdx = ad$, $d \neq 0$. We take b to be positive, as we can always take care of negative fractions by a negative a. We can now let the equations count themselves by the trick of thinking of the integers I written to base 13. We let the character for 10 in the base 13 system be x, the character for 11 be $=$, and the character for 12 be $-$. Then every positive integer has exactly one form to base 13; for example,

I	I_{13}		I	I_{13}
8	8		14	11
9	9		25	1$-$
10	x		26	20
11	$=$		100	79
12	$-$		169	100
13	10			

$$10 \cdot 13^2 + 12 \cdot 13 + 1 \qquad x - 1$$

Now with a and b written to base 10, each equation looks like an integer to base 13; for example, $2x = -3$ would be

$$2 \cdot 13^4 + 10 \cdot 13^3 + 11 \cdot 13^2 + 12 \cdot 13 + 3 = 81110.$$

We use the order of the positive integers in counting the equations. If we tried to count first all the equations $1x = a$, $a = 1, 2, 3, \ldots$, we could count forever without covering other values of b. The correspondence we have just developed shows that the number of all the equations, and hence the number of elements of Q, is less than or equal to $\aleph_0$, the number of elements of I.

Now since $q \leq \aleph_0$ and also $q \geq \aleph_0$, $q = \aleph_0$.

R HAS $2^{\aleph_0}$ MEMBERS

Since R contains all decimal numbers, it contains among others the terminating and repeating ones that represent members of Q (or we can represent these in their nonterminating form as in Definition 2–16). From this we know that R contains at least $\aleph_0$ members.

We show that the real numbers r with $0 < r < 1$ are not countable, in a proof by contradiction.

Suppose the real numbers between 0 and 1 are countable. This means that there is a one-to-one correspondence between the real numbers and the natural numbers, so that there is a first, r_1 (not necessarily smallest, just the one that corresponds to 1), a second, r_2, a third, r_3, and so on. We can then construct a new real number r between 0 and 1 that had been left out of the correspondence: Look at the first digit to the right of the decimal point in r_1; if it is a 1, let the first digit of r be 2; otherwise let the first digit of r be 1. Look at the second digit of r_2; if it is a 1, let the second digit of r be 2; otherwise let the second digit of r be 1. Continue to determine the nth digit of r from r_n as 2 if r_n has 1 in the nth place, as 1 if the nth digit of r_n is not 1. The new r, then, was missed in the supposed correspondence between the part of R between 0 and 1 and N, for it differs from every r_n. We show how this might look in a special case.

$$r_1 = 0.013412 \quad r = 0.1^{\cdot} \qquad\qquad \text{or } r = 0.121112\cdots$$
$$r_2 = 0.312189 \qquad\qquad 2$$
$$r_3 = 0.959111 \qquad\qquad 1$$
$$r_4 = 0.090912 \qquad\qquad\quad 1$$
$$r_5 = 0.881238 \qquad\qquad\qquad 1$$
$$r_6 = 0.012341 \qquad\qquad\qquad 2$$

$$\cdots$$

$r \neq r_1$ nor r_2 nor r_3 nor r_4 nor r_5 nor r_6.

Then R has more than $\aleph_0$ members. Its cardinal number (its "count") is taken to be $2^{\aleph_0}$, which is consistent with the fact that there are 2^x numbers whose binary representations have length x. (For $x = 3$ we would have 000, 001, 010, 011, 100, 101, 110, 111.) Since real numbers may have representations of countable length $\aleph_0$, we say that R has $2^{\aleph_0}$ members. Notice that the nonfractional part of each real number is an integer, a member of I, which has only $\aleph_0$ members.

With an appreciation of mathematical uncountability, you may be amused to note the subjective range in expressions like "the parade passed with its innumerable marchers" or "he kissed her countless times."

C HAS $2^{\aleph_0}$ MEMBERS

One correspondence between the members r or R in the range $0 < r \le 1$ and the members $s + ti$ of C in the range $0 < s \le 1, 0 < t \le 1$ calls for splitting the nonterminating decimal representation of r (Definition 2–16) into blocks, beginning at the decimal point and ending each block with the first nonzero digit occurring after the previous block. For instance, break

$$0.1230040000506 \cdots$$

into 1, 2, 3, 004, 00005, 06, ... Make up the decimal representation of an s and a t from these blocks, alternating, using the odd-numbered ones for s, even for t, as

$$s = 0.1300005 \cdots, \quad t = 0.200406 \cdots.$$

This process can be reversed, so that the correspondence is indeed one-to-one.

Exercise C–1. Under the reverse process, what real number would correspond to $0.012003045608 \cdots + (i)0.3330706098 \cdots$?

To show why we do not use the simpler correspondence of 1-digit blocks no matter where the 0's lie, notice that the number corresponding to the number

$$0.1020302050604070 \cdots$$

having 0's in all even-numbered decimal places would fail to lie in the desired range, since we require $t > 0$.

The procedure shown here for unit intervals can be extended to a correspondence between all of R and all of C.

GRAPHING COMPLEX NUMBERS

The two components of a complex number can be plotted as a point on a Cartesian plane with the real part as abscissa, imaginary part as ordinate. The points of the plane, or the vectors from the origin to the points, form convenient geometric representations of the complex numbers. The plane of complex numbers is called the "Argand" plane after one of its inventors, although both Gauss and Caspar Wessel used such planes before Argand did.

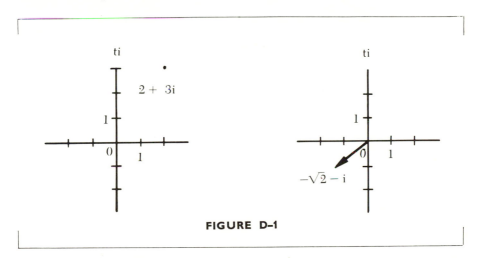

FIGURE D–1

Exercise D–1. Plot in the Argand plane $-2.\overline{9} + 7i$, 7, and $3 - 7i$. Describe, as to a student, exactly how to plot $s + ti$.

Exercise D–2. Plot as vectors from the origin $4 + i$, $2 + 6i$, and their sum, $6 + 7i$. Plot $3 + 3i$, $-3 + 3i$, and their sum.

Think of the complex number $s + ti$ plotted as a vector from the origin. Its length, from the Pythagorean theorem, is $r = \sqrt{s^2 + t^2}$. It makes an angle θ with the positive real axis.

In trigonometry the function s/r is given the name cosine θ (cos θ) and is studied as a function of the angle θ. Similarly, sine θ, abbreviated sin θ, is the function t/r, and tangent θ, or tan θ, is t/s. The length r is taken to be the absolute value of the length and so is always thought of as positive, but s or t or both may be negative in these quotients.

We have $s = r \cos \theta$ and $t = r \sin \theta$, so we can write our complex number

$$s + ti = r(\cos \theta + i \sin \theta). \tag{1}$$

167

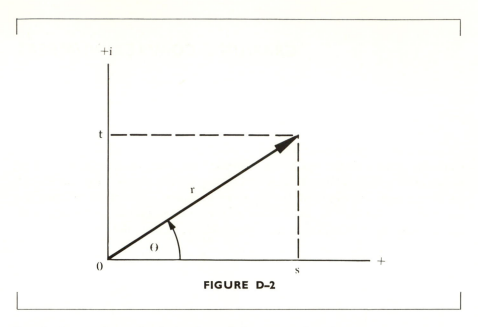

FIGURE D-2

In calculus, we have the derivatives of elementary functions:

$$\frac{d(\cos\theta)}{d\theta} = -\sin\theta; \quad \frac{d(\sin\theta)}{d\theta} = \cos\theta; \quad \frac{d(e^{i\theta})}{d\theta} = ie^{i\theta}. \tag{2}$$

Using these in (1), thought of as a function of θ, we find

$$\frac{d[r(\cos\theta + i\sin\theta)]}{d\theta} = r(-\sin\theta + i\cos\theta) = ir(\cos\theta + i\sin\theta),$$

so that the derivative of the function (1) equals the function times i. Comparing with the derivative of $e^{i\theta}$ in (2), we now write the complex number in terms of its length and angle as:

$$re^{i\theta} = r(\cos\theta + i\sin\theta) = r\left(\frac{s}{r} + \frac{it}{r}\right) = s + ti. \tag{3}$$

What happens to the vector representation when we add or multiply complex numbers? Addition is easiest to picture using the $s + ti$ form. The sum vector forms the diagonal of the parallelogram formed with the two given vectors as adjacent sides.

Exercise D-3. Verify the parallelogram law in Exercise D-2. In Figure D-3 how long is OA? BC? OD? EF?

It is easier to multiply two vectors in the form $re^{i\theta}$. We have

$$r_1e^{i\theta} \cdot r_2e^{i\sigma} = r_1r_2e^{i(\theta+\sigma)},$$

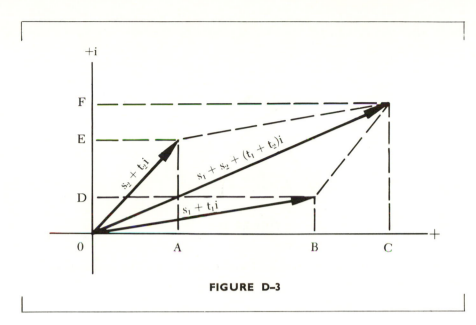

FIGURE D-3

showing that the product vector has as length the product of the two given lengths and as angle the sum of the given angles.

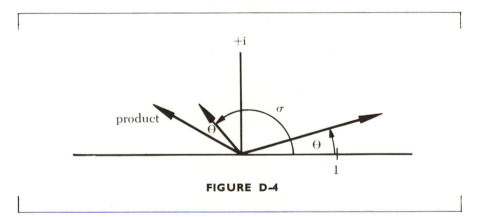

FIGURE D-4

Exercise D-4. Find the products algebraically and compare with the results obtained graphically: (*a*) $6 + i$ and $-1 + 5i$, (*b*) i and i, (*c*) the cube $[(-1 + i\sqrt{3})/2]^3$. Show both the second and the third powers on the vector diagram.

Angles can be measured in degrees, in which case a straight angle is a $180°$ angle, and $360°$ represents one complete rotation about the origin. Angles can also be measured in radians, a radian being defined as the angle that cuts off one unit of arc length in the unit circle. Since the whole circumference of a circle of radius r is $2\pi r$, we see from the unit circle with $r = 1$ that one complete rotation about the origin is equivalent to 2π radians, so that $2\pi = 360°$. In Exercise D-4 (*c*) we can obtain a hint of how to construct roots of unity in the

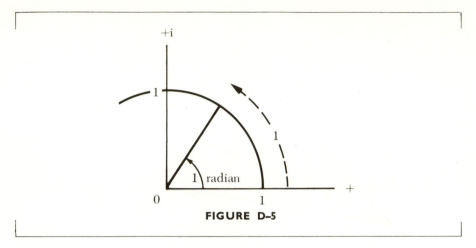

FIGURE D–5

complex plane. The square roots of 1 are $+1$ and -1. The fourth roots of 1 are i, -1, $-i$, and $+1$.

   –Exercise D–5.   Verify algebraically and graphically that $u^4 = 1$ for $u = i$, -1, $-i$, and $+1$.

   The nth roots of unity can be found around the unit circle at angles $\theta = k2\pi/n$, $k = 0, 1, 2, \ldots, n - 1$. Then in the form $re^{i\theta}$ we have the roots $e^{ik2\pi/n}$ (cf. Definition 9–1 for the nth roots of unity).

   Exercise D–6.   Verify algebraically that the nth power of $e^{ik2\pi/n}$, $k = 0, 1, \ldots, n - 1$, is 1.

   Exercise D–7.   Draw separate graphs of the fifth, sixth, and seventh roots of unity and form a star or other geometric design based on one of them.

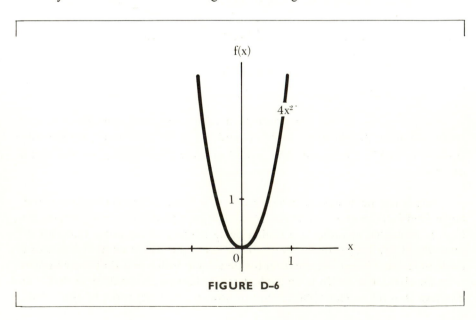

FIGURE D–6

When we graph in the real plane a polynomial function of x such as $4x^2$, we obtain no negative function values and have, therefore, no points plotted below the x-axis.

If we include in addition to the real plane an "Argand" plane for complex values (usually drawn perpendicular to the real plane), we can allow for negative

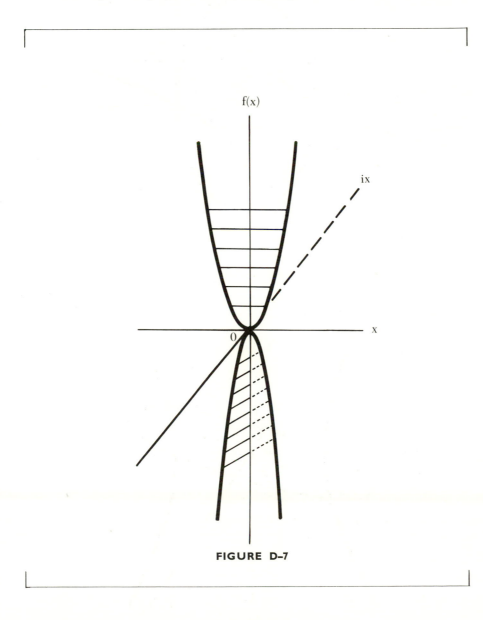

FIGURE D-7

function values. For instance, we can plot for the polynomial $f(x) = 4x^2$, the number pair $(2i, -16)$ as a point in the plane of the $f(x)$-axis $x = 0$ (because the real part of $2i$ is zero), 2 units along the positive ix axis, and 16 units down parallel to the $f(x)$-axis.

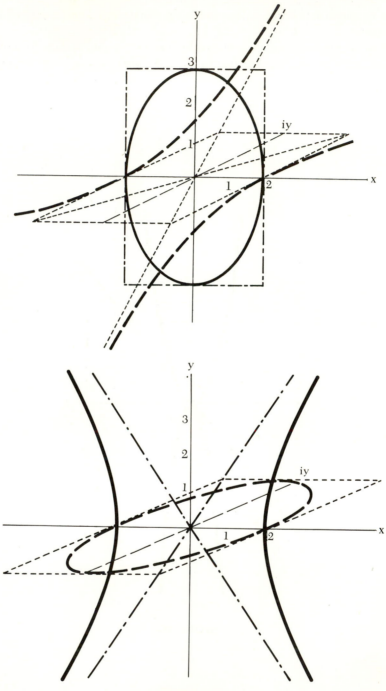

FIGURE D–8

If the ellipse $x^2/2^2 + y^2/3^2 = 1$ and the hyperbola $x^2/2^2 - y^2/3^2 = 1$ are plotted with both a real plane and a complex plane, they appear as complements of each other, for we can replace $-y^2$ by $(iy)^2$.

A LATTICE OF ABSTRACT ALGEBRAS

Definition E-1. A **lattice** is a partially ordered set with two operations **"least upper bound, *a* ∨ *b*"** and **"greatest lower bound *a* ∧ *b*."** (Without defining a "partial ordering," we merely state that in our examples "is not lower than" and "contains or equals" meet the requirements.)

The diagram of Figure 2–6 forms an example of a lattice if we take *a* ∨ *b* for two points *a* and *b* to be the lowest point not lower than *a* or *b*. Symmetrically, take *a* ∧ *b* to be the highest point not higher than *a* or *b*. Then, for instance, Groups ∨ Rings would be Groups, and Rings with unit element ∧ Commutative rings would be Commutative rings with unit element.

Similarly, the system of abstract algebras with the inclusions symbolized in Figure 2–6 form a lattice, with *A* ∨ *B* for any two types of algebras being the minimal type that contains both (equality permitted), *A* ∧ *B* the maximal type contained in both (equality permitted).

A lattice is itself an abstract algebra, so we have shown a set of abstract algebras whose comparisons form an abstract algebra.

Perhaps you can sense even from this short introduction to the subject what a powerful tool we have here for thinking about thinking.

MODULAR ARITHMETIC

Scattered among the exercises in this book is material on the integers modulo an integer $n > 1$, $I/\langle n \rangle$. Collect this information in a report, studying $I/\langle n \rangle$ in its own right instead of as illustrative material. Such a report should include proofs of what kind of abstract algebra $I/\langle n \rangle$ represents if n is prime and if n is composite, examples of "clock arithmetic" or other modular systems. See Exercises 3–47, 4–8, 5–14, and 5–15.

Lemma 7–2 shows that an irreducible polynomial over a field of characteristic zero has no multiple zeros, that is, no factor $(x - \theta)^m$ for $m > 1$. Show that $x^2 - z$ is irreducible in $T(z)$, where T is the field of integers modulo 2, and $T(z)$ the field of rational combinations of z and elements of T. However, show that if θ is a zero of $x^2 - z$ in an extension field U, then $(x - \theta)^2$ divides $x^2 - z$ in U.

GRAPHICAL GROUP
REPRESENTATIONS

Look up the paper "Graphical Group Representations" by J. E. Maxfield in "Mathematics Magazine," Jan–Feb, Vol. 27, No. 3, 1954, pages 169–174. If your library cannot supply this reference, you can probably borrow the volume through inter-library loan, a useful process to explore, anyway. You may need some help from your instructor to adapt the notation of the paper to suit the notation of this book, but the main illustration in the paper concerns the quaternion group. Compare with our Figure 3–9. You may prepare a report on the paper, or construct a large color diagram of the quaternion group. See whether you can follow the ideas of the paper enough to draw a color diagram for some other group, such as the four-group.

HISTORY OF MATHEMATICS

Our "Historical Intermission" gives you an introduction to mathematical history. There are several ways you might want to follow this up:

1. Study the time in which Galois and Abel worked, from whatever angle interests you especially.

2. Read some of E. T. Bell's *Men of Mathematics*, Dover, 1937, watching especially for references to other mathematicians we have mentioned: Gauss, Cauchy, Hermite, and others.

3. Read biographies of mathematicians, investigating in particular how each was trained, what kind of teaching seemed best for him, and so forth.

4. Prepare some notes that you might use to interest an anti-science student in algebra, through an appeal to interest in history.

MORE ABOUT FIELD EXTENSIONS

Theorem I–1. Let $p(x)$ be an irreducible polynomial in $F[x]$, and let θ and $\bar{\theta}$ be conjugate zeros of $p(x)$ in $E \supseteq F$. Then $F(\theta) \cong F(\bar{\theta})$. In fact, let σ be an isomorphism between F and $\bar{F}$, and let $\bar{\theta}$ be a root of the polynomial $p^\sigma \in \bar{F}[x]$ corresponding to p under the isomorphism. Then σ can be extended to an isomorphism between $F(\theta)$ and $\bar{F}(\bar{\theta})$ in which θ corresponds to $\bar{\theta}$.

PROOF: By two applications of Theorem 7–3*iii*, we find $F(\theta) \cong F[x]/\langle p(x) \rangle$ and also $\bar{F}(\bar{\theta}) \cong \bar{F}[x]/\langle p^\sigma(x) \rangle$. Let these two isomorphisms be designated by σ_2 and σ_3, respectively. Then $F(\theta) \cong \bar{F}(\bar{\theta})$, where $f \in F(\theta)$ corresponds to $\sigma_3^{-1}\sigma\sigma_2 f$. The special case $F = \bar{F}$ yields the result for conjugates. The isomorphism fixes the elements of F. ∎

Notice that when we construct $F(\theta) \cong F[x]/\langle p(x) \rangle$, where θ is a zero of $p(x)$ in E, then a second zero, $\bar{\theta}$, of $p(x)$ in E may or may not lie in the constructed field. If $\bar{\theta} \in F(\theta)$, then $F(\theta, \bar{\theta}) = F(\theta)$, and $\bar{\theta}$ has already been effectively adjoined. If not, we can extend $F(\theta)$ by constructing $F(\theta)[x]/\langle q(x) \rangle$, where $q(x)$ is an irreducible factor of $p(x)$ in $F(\theta)[x]$ having $\bar{\theta}$ as a zero. This extension does contain $\bar{\theta}$ (Theorem 7–2) and in it we can form both $F(\theta)$ and $F(\bar{\theta})$, isomorphic by Theorem I–1, showing that we would have essentially the same field by adjoining $\bar{\theta}$ first.

Exercise I–1. Let F be the field of rationals Q, and let $p(x)$ be the irreducible polynomial $x^2 + 1$, which has conjugate zeros i and $-i$ in an extension of Q. Show that $Q(i) \cong Q(-i)$ under $+$ and $\cdot$.

Lemma 7–2. Let F be any field, and let $f(x)$ be a polynomial of degree d in $F[x]$. Then $f(x)$ has no more than d zeros in any extension field $E \supseteq F$.

Let F be a field of characteristic zero, and let $p(x)$ of degree n be a monic irreducible polynomial in $F[x]$. If $(x - \theta)^m$, with $m > 0$, divides $p(x)$ in $E[x]$, where E is an extension of F, then $m = 1$, so that the zeros of $p(x)$ in E are distinct.

Let $p(x) \in Q[x]$ be $x^3 + 2x + 2$. In E let $(x - \theta)^2$ divide $p(x)$. Then since $x - \theta$ divides $p(x)$, the remainder upon division is zero. We find the quotient $x^2 + \theta x + (\theta^2 + 2)$.

$$
\begin{array}{r}
x^2 + \theta x + (\theta^2 + 2) \\
x - \theta \overline{)x^3 \qquad\qquad\quad + 2x\ + 2} \\
\underline{x^3 - \theta x^2} \qquad\qquad\qquad\quad \\
\theta x^2 \qquad\quad + 2x \qquad \\
\underline{\theta x^2 \qquad\quad - \theta^2 x} \qquad \\
(\theta^2 + 2)x\ + 2 \\
\underline{(\theta^2 + 2)x - \theta^3 - 2\theta} \\
\theta^3 + 2\theta + 2 = p(\theta) = 0
\end{array}
$$

Is θ a zero of $q(x) = x^2 + \theta x + (\theta^2 + 2)$? If so, it is a zero of $x^2 + x^2 + (x^2 + 2)$, or $3x^2 + 2$, a polynomial of degree 2. (We recognize $3x^2 + 2$ as the derivative of $x^3 + 2x + 2$.)

FIGURE I–1

PROOF: We count θ as a zero of $f(x)$ m times (**multiplicity m**) if $(x - \theta)^m$ divides $f(x)$ in $E[x]$ but $(x - \theta)^{m+1} \nmid f(x)$. By the Factor Theorem 6–4, if θ is a zero of multiplicity m of $f(x)$, then $f(x) = (x - \theta)^m g(x)$ for some polynomial $g(x)$ and the degree of $g(x)$ is $d - m$, from polynomial multiplication. If $\theta_2 \neq \theta$ is also a zero of $f(x)$, then since $x - \theta_2$ is relatively prime to $(x - \theta)^m$, we have from Theorem 6–9, $x - \theta_2 \,|\, g(x)$. Continuing in this way, we can show that, counting each zero as many times as its multiplicity, $f(x)$ has no more zeros than its degree in any extension field E.

Now let F be a field of characteristic zero and let $p(x)$ of degree n be a monic irreducible polynomial in $F[x]$. (Follow the illustration in Figure I–1.)

Since $(x - \theta)$ divides $p(x)$ in $E[x]$, there is a quotient $q(x) \in E[x]$ for which $p(x) = (x - \theta)q(x)$. We have $q(x) = \sum_{i=0}^{n-2} r_i(\theta)x^i + x^{n-1}$, where each coefficient $r_i(\theta)$ is a polynomial in θ of degree $\leq n - 1 - i$, as we could prove inductively, following the long division process illustrated in Figure I–1. Then the degree of $q(\theta)$ as a polynomial in θ and x is $\leq n - 1$. In fact, we could prove that $q(\theta)$ for $x = \theta$ is the derivative $p'(\theta)$ with highest degree term $n\theta^{n-1}$. Then if $q(\theta) = 0$, θ is a zero of a polynomial of positive degree less than n, contrary to Theorem 7–1. ∎

Theorem I–2. Let $f(x)$ be a polynomial in $F[x]$, where F is a field. If A is a splitting field for $f(x)$ over F and B is a splitting field over an isomorphic field $\bar{F} \cong F$ for $\bar{f}(x)$, the corresponding polynomial, then A is isomorphic to B. In particular, any two splitting fields for $f(x)$ over F are isomorphic.

PROOF: In $A[x]$, $f(x)$ has n linear factors $x - \theta_i$, $i = 1, 2, \ldots, n$. By Theorem 7–5, $F(\theta_1, \theta_2, \ldots, \theta_n)$ is a splitting field for $f(x)$ over F, and it is intermediate between F and A since the θ's are in A. Then since the splitting field A has a minimal property by definition, we have $F(\theta_1, \theta_2, \ldots, \theta_n) = A$. Similarly, the splitting field $B = \bar{F}(\phi_1, \phi_2, \ldots, \phi_n)$, where $\bar{f}(x)$ has the factors $x - \phi_i$, $i = 1, 2, \ldots, n$, in $B[x]$. (By Lemma 7–2 $f(x)$ has no more zeros than its degree, n.)

Let $f(x) = p(x)g(x)$, where $p(x)$ is an irreducible factor of $f(x)$ in $F[x]$ of degree $m \leq n$. If all factors of $f(x)$ in $F[x]$ have degree < 2, then $f(x)$ splits in F and F is its splitting field over F. Then suppose $m > 1$. Then from $f(x) = p(x)g(x) = (x - \theta_1)(x - \theta_2) \cdots (x - \theta_n)$ in $A[x]$, and from Theorem 6–9 applied to $A[x]$, we see that $p(x)$ has m zeros among the θ's in A. Similarly, $\bar{p}(x)$ has m zeros among the ϕ's in B. From Theorem I–1 it makes no difference, up to an isomorphism, which of the θ's we adjoin first to F. Form $F(\theta_1) \cong F[x]/\langle p(x) \rangle$, and form $\bar{F}(\phi_1) \cong \bar{F}[x]/\langle \bar{p}(x) \rangle$. In $F(\theta_1)$, $p(x)$ has the linear factor $x - \theta_1$, and the powers $1, \theta_1, \theta_1^2, \ldots, \theta_1^{m-1}$ form a linear basis for the field. Similarly, $1, \phi, \phi^2, \ldots, \phi^{m-1}$ form a basis for $\bar{F}(\phi_1)$. Using the given isomorphism between F and $\bar{F}$, extend the correspondence so that θ_1 and ϕ_1 correspond. Then every element of $F(\theta_1)$ can be expressed linearly over F in terms of the θ_1 basis and so converted to a corresponding element expressed linearly in terms of the ϕ_1 basis in $\bar{F}(\phi_1)$.

If another zero, say θ_2, lies in $F(\theta_1)$, it is already effectively adjoined and $F(\theta_1, \theta_2) = F(\theta_1)$. In this case one of the ϕ's, say ϕ_2, must lie in $\bar{F}(\phi_1)$, since the selection of θ_1 or ϕ_1 was purely formal in the construction of $F[x]/\langle p(x) \rangle$. Let $A_1 = F(\theta_1) = F(\theta_1, \theta_2, \ldots, \theta_k)$ and suppose no other θ_i for $i > k$ lies in A_1. Then

$$q_{A_1}(x) = p(x)/[(x - \theta_1)(x - \theta_2) \cdots (x - \theta_k)]$$

is a quotient polynomial in $F(\theta_1)$ and has its corresponding quotient polynomial $q_{B_1}(x)$ in $B_1 = \bar{F}(\phi_1) = \bar{F}(\phi_1, \phi_2, \ldots, \phi_k)$. (Here the ϕ's may have needed renumbering.) The corresponding quotient polynomials have corresponding irreducible factors $q_{A_1}^*(x)$ in $A_1[x]$ and $\bar{p}_{B_1}^*(x)$ in $B_1[x]$, respectively. We follow again the construction of Theorem 7–2, obtaining $A_2 = A_1(\theta_{k+1}) \cong A_1[x]/\langle q_{A_1}^*(x) \rangle$ and $B_2 = B_1(\phi_{k+1}) \cong B_1[x]/\langle q_{B_1}^*(x) \rangle$. We extend the isomorphism between A_1 and B_1 to an isomorphism between A_2 and B_2 with θ_{k+1} and ϕ_{k+1} in correspondence (Theorem I–1).

From Lemma 7–2 there are no more zeros for the irreducible factor $p(x)$ than the degree of $p(x)$. We continue the construction, at each stage extending the isomorphism, until at some stage all n of the zeros have been adjoined and the resulting extensions $F(\theta_1, \theta_2, \ldots, \theta_n) = A$ and $\bar{F}(\phi_1, \phi_2, \ldots, \phi_n) = B$ are isomorphic. ∎

Exercise I–2. Factor $f(x) = x^4 + 3x^2 + 2$ into two irreducible quadratic factors $a(x)b(x)$ in $Q[x]$. Follow the proof of Theorem I–2, letting A_1 be isomorphic to $Q[x]/\langle a(x) \rangle$ and B_1 be isomorphic to $Q[x]/\langle b(x) \rangle$. Obtain the splitting fields, and show that they are isomorphic.

Exercise I–3. Let F be the rational field Q. Let $p(x)$ be the quadratic polynomial $ax^2 + bx + c$, $a \neq 0$. Show that the splitting field for this polynomial over Q is either Q itself or $Q(\theta)$, where θ is a zero of the polynomial in $Q[x]/\langle ax^2 + bx + c \rangle$; that is, show that $\bar{\theta}$ is in $Q(\theta)$.

Exercise I–4. Let $f(x) = x^3 + 3x^2 + 4$ in $Q[x]$. The function has a zero $\theta_1 = a + b - 1$, where $a = (-3 + \sqrt{8})^{1/3}$ and $b = (-3 - \sqrt{8})^{1/3}$ in the real field. Show by graphing the function and studying its derivative, changes of sign, and so on, that $f(x)$ fails to split in $Q(\theta_1)$. Use Exercise I–3 to show that $f(x)$ splits in $Q(\theta_1, \theta_2)$, where θ_2 is another zero of $f(x)$. Verify that θ_1 is a zero of $f(x)$, noting that $ab = 1$. Use synthetic division to find the quotient $f(x)/(x - \theta_1)$, and find θ_2 and θ_3 in terms of θ_1 from the quadratic formula.

> **Theorem I–3.** Let $F \subseteq E \subseteq D$ be fields. If $[D:E] = m$ and $[E:F] = n$, then $[D:F] = mn$.

PROOF: Let s_i, $i = 1, 2, \ldots, n$, be a linear basis for E over F so that each $e_j \in E$ has an expression

$$e_j = \sum_{i=1}^{n} f_{ij} s_i, \quad f_{ij} \in F.$$

Let t_j, $j = 1, 2, \ldots, m$, be a linear basis for D over E, so that each $d \in D$ has an expression

$$d = \sum_{j=1}^{m} e_j t_j, \quad e_j \in E.$$

Substituting for each coefficient $e_j \in E$ its linear form over F, we have

$$d = \sum_{j=1}^{m} \sum_{i=1}^{n} f_{ij}(s_i t_j).$$

The mn products $s_i t_j$, $i = 1, 2, \ldots, n$, $j = 1, 2, \ldots, m$, all lie in D and are linearly independent over F (as indicated in the exercises to follow), so they constitute a basis for D over F. ∎

Exercise I–5. Let s_1 and s_2 be a basis for E over F, and let t_1 and t_2 be a basis for D over E. Show that $s_1 t_1$, $s_1 t_2$, $s_2 t_1$, and $s_2 t_2$ are linearly independent over F by supposing that

$$f_{11} s_1 t_1 + f_{12} s_1 t_2 + f_{21} s_2 t_1 + f_{22} s_2 t_2 = 0$$

with not all of the f's equal to zero and showing that this would imply a dependence among the s's or the t's.

Exercise I-6.   Describe how Exercise I–5 might be generalized to complete the proof of Theorem I–3.

Definition I-1.   Let $F \subseteq E$ be fields. E is a **simple extension** of F if, for some element $\theta \in E$, E equals $F(\theta)$. Then θ is a **primitive element** of E over F.

Theorem I-4.   Let F be a field of characteristic zero, and let $\theta_1, \theta_2, \ldots, \theta_k \in E \supseteq F$ be algebraic relative to F. Then $F(\theta_1, \theta_2, \ldots, \theta_k)$ is a simple algebraic extension of F with a primitive element ψ.

PROOF: We prove the theorem for $k = 2$, which will enable us to conclude it for a general k, since the θ's can be adjoined one at a time to the preceding extension. Let θ and ϕ be two elements of E, algebraic over F. If ϕ is an element of $F(\theta)$, then $F(\theta, \phi) = F(\theta)$ is a simple extension with θ as a primitive element. Then suppose ϕ is not in $F(\theta)$.

Since θ is algebraic relative to F, we know from Theorem 7–1 that it is a zero of an irreducible polynomial $p(x)$ in $F[x]$. Similarly, ϕ is a zero of $q(x)$ in $F[x]$. Let θ and its conjugates be $\theta = \theta_1, \theta_2, \ldots, \theta_m$. Let ϕ and its conjugates be $\phi = \phi_1, \phi_2, \ldots, \phi_n$. From Lemma 7–2 the conjugates in each case are distinct. Select $f \in F$ not equal to any of the ratios $(\theta_i - \theta_h)/(\phi_k - \phi_j)$ and not equal to zero; that is, in the infinite field F avoid this finite number of elements.

Let $\psi = \theta + f\phi$. We shall show that $F(\psi) = F(\theta, \phi)$. First we note that ψ lies in $F(\theta, \phi)$, for it is a linear expression in θ and ϕ, coefficients 1 and f in F. We are going to prove that ϕ lies in $F(\psi)$. (Follow the illustration in Figure I–2.) Form a polynomial function of x in $F(\psi)[x]$, $s(x) = p(\psi - fx)$. When x takes on the value ϕ, then $\psi - fx = \psi - f\phi = \theta$, so $s(\phi) = p(\theta) = 0$.

Then $s(x)$ and $q(x)$ are two polynomials in $F(\psi)[x]$ having the common zero ϕ. Let their g.c.d. in $F(\psi)[x]$ be $ct(x)$, where $c \in F$ and $t(x)$ is monic. We want to prove that $t(x) = x - \phi$. From the linear expression

$$t(x) = u(x)s(x) + v(x)q(x),$$

we see that in the extension $F(\psi, \phi)$

$$t(\phi) = u(\phi)s(\phi) + v(\phi)q(\phi) = 0 + 0 = 0,$$

so that $t(\phi) = 0$. Then $\deg t(x) > 0$ and $(x - \phi) \,|\, t(x)$ in $F(\psi, \phi)$. Every factor of $t(x)$ must be a factor of $q(x)$ and of $s(x)$. But all the factors of $q(x)$ in an extension field in which $q(x)$ splits are $x - \phi_j$, $j = 1, 2, \ldots, n$. Suppose $x - \phi_j$ divides $s(x)$ with the conjugate $\phi_j \neq \phi = \phi_1$. Then ϕ_j is a zero of $s(x)$, so that $\psi - f\phi_j$ is a zero of $p(x)$, that is, one of the conjugates θ_i. We would have two expressions for ψ, then, which would be equal: $\psi = \theta_i + f\phi_j = \theta_1 + f\phi_1$. From this f would equal $(\theta_i - \theta_1)/(\phi_1 - \phi_j)$, contrary to its selection, so no conjugate $\phi_j \neq \phi$ can be zero of $t(x)$. Then all the factors of $t(x)$ are $x - \phi$. But from Lemma 7–2 $q(x)$ has no multiple zeros, so $t(x) = x - \phi$.

Let $F = Q$, $\theta = \sqrt{2}$, $\phi = i = \sqrt{-1}$. Form $\psi = \sqrt{2} + fi$.

$$\theta = \theta_1 = \sqrt{2}, \quad \theta_2 = -\sqrt{2}; \quad \phi = \phi_1 = i, \quad \phi_2 = -i$$

$$p(x) = x^2 - 2, \quad q(x) = x^2 + 1$$

$$s(x) = p(\psi - fx) = (\psi - fx)^2 - 2 = f^2x^2 - 2f\psi x + \psi^2 - 2$$

$$ct(x) = (s(x), q(x)).$$

Use the Euclidean algorithm to find $ct(x)$.

$$
\begin{array}{r}
f^2 \\
x^2 + 1 \overline{\smash{)}\, f^2x^2 - 2f\psi x + \psi^2 - 2} \\
\underline{f^2x^2 + f^2} \\
- 2f\psi x + \psi^2 - f^2 - 2
\end{array}
$$

Then

$$-2f\psi\left(x - \frac{\psi^2 - f^2 - 2}{2f\psi}\right) = s(x) - f^2q(x).$$

When $x = i$, $s(i) = p(\sqrt{2}) = 0$ and $q(i) = 0$, so

$$\phi - \frac{\psi^2 - f^2 - 2}{2f\psi} = 0,$$

or

$$\phi = \frac{\psi^2 - f^2 - 2}{2f\psi} \in F(\psi).$$

To show that this really gives us a formula for i in terms of $\sqrt{2} + fi$, compute

$$\frac{\psi^2 - f^2 - 2}{2f\psi} = \frac{(\sqrt{2} + fi)^2 - f^2 - 2}{2f(\sqrt{2} + fi)} = i.$$

FIGURE I–2

Then $\phi = -t(0) = -u(0)s(0) - v(0)q(0) \in F(\psi)$. From $\theta = \psi - f\phi$, we have $\theta \in F(\psi)$. Then $F(\psi) = F(\theta, \phi)$. From Theorem 7–4 $F(\psi)$ is an algebraic extension of F. ∎

In Theorem I–4 we assumed that $\theta_1, \theta_2, \ldots, \theta_k$ are algebraic relative to F. In the following corollary we are able to prove that basis elements for any finite extension are algebraic relative to F and to conclude, then, that every finite extension is an algebraic extension.

Corollary. Let $[E:F] = n$, where F has characteristic zero. Then E is a simple algebraic extension of F.

PROOF: From the definition of $[E:F]$, E has a linear basis of n elements $\theta_i \in E$, $i = 1, 2, \ldots, n$, over F. Then from Theorem 7–3ii, $E \subseteq F(\theta_1, \theta_2, \ldots, \theta_n)$, for each element $\sum_{i=1}^{n} f_i\theta_i, f_i \in F$, of E is a rational combination of the θ's. Since the field E contains all quotients with nonzero denominators, we have $E = F(\theta_1, \theta_2, \ldots, \theta_n)$.

We can show that each basis element $\theta = \theta_i$ for some $i = 1, 2, \ldots,$ or n is algebraic relative to F: Since each power of θ can be expressed in terms of the n basis elements, the $n + 1$ powers $1, \theta, \theta^2, \ldots, \theta^n$ have a linear dependence $\sum_{j=0}^{n} f_i\theta^i = 0$. Then θ is a zero of $\sum_{j=0}^{n} f_i x^i \in F[x]$ and so is algebraic relative to F. From Theorem I–4, $F(\theta_1, \theta_2, \ldots, \theta_n)$ is a simple algebraic extension of F. ∎

MORE ABOUT GALOIS THEORY

In the following theorem we arrive at a second necessary and sufficient condition for a finite extension of a field of characteristic zero to be a normal extension. The first such condition we used to define a normal extension (Definition 8–4). It was that any irreducible polynomial with one zero in the extension split in the extension.

Theorem J–1. Let E be a finite extension of a field F of characteristic zero. Then E is a normal extension of F if and only if F is the fixed field of $G(E/F)$.

PROOF: Notice that $G(E/F)$ is made up of the automorphisms of E that fix F, but they may fix other elements as well. If the fixed field of $G(E/F)$ is K, we have $F \subseteq K$, and the object of this proof is the result: $F = K$ if and only if E is normal.

First we take care of the case $E = F$. In this case each irreducible polynomial in $F[x]$ with a zero in $E = F$ is a linear polynomial and splits in F, so F is normal over F, and F is the fixed field of $G(F/F) = \{\hat{1}\}$.

Then suppose $E \neq F$. From Theorem I–4 the finite extension E is a simple extension with primitive element ψ. From Theorem 7–4 $E = F(\psi)$ has a basis $1, \psi, \ldots, \psi^{n-1}$, where ψ is a zero of an irreducible polynomial $p(x)$ of degree $n = [E:F]$.

If E is normal it contains the conjugates $\psi = \psi_1, \psi_2, \ldots, \psi_n$ of ψ. If $e \in E$ is fixed by $G(E/F)$, then

$$e = \sum_{i=0}^{n-1} f_i \psi^i = \sum_{i=0}^{n-1} f_i \psi_j^i, \tag{1}$$

for each $j = 1, 2, \ldots, n$. If we let $f(x)$ be the polynomial

$$f(x) = -e + \sum_{i=0}^{n-1} f_i x^i,$$

then (1) says that the polynomial $f(x)$ of degree $n - 1$ has n zeros $\psi_1, \psi_2, \ldots, \psi_n$. But this cannot occur unless $f(x)$ is the zero polynomial (Lemma 7–2), so that $f_i = 0$ for $i > 0$ and $-e + f_0 = 0$. Thus $e = f_0 \in F$, showing that if E is normal, then each element fixed by $G(E/F)$ is in F.

Conversely, suppose F is the fixed field of $G(E/F)$. Let $q(x)$ be any irreducible polynomial of degree d in $F[x]$ with a zero $\theta = \theta_1$ in E. If $\theta \in F$, then

$d = 1$ and $q(x) = x - \theta$ splits in E. If $d > 1$, then $\theta \notin F$, since $q(x)$ is irreducible over F. Then, since F is the fixed field of $G(E/F)$, there is an automorphism $\sigma_1 \in G(E/F)$ that moves θ_1; that is, $\sigma_1\theta_1 \neq \theta_1$. From

$$q(\sigma_1\theta_1) = \sigma_1 q(\theta_1) = \sigma_1(0) = 0,$$

we have

$$\sigma_1\theta_1 = \theta_2,$$

a conjugate of θ. Because σ_1 is an automorphism of E, $\theta_2 \in E$.

Now proceed inductively to prove that all d zeros of $q(x)$ lie in E: Let

$$r(x) = (x - \theta_1)(x - \theta_2) \cdots (x - \theta_k),$$

$2 \leq k \leq d$, θ's in E, and let $q(x) = r(x)s(x)$, where $s(x) \in E[x]$ of degree ≥ 0. If $\deg s(x) = 0$, then $q(x) = r(x)s$ splits in E. If $\deg s(x) > 0$, then $r(x) \notin F[x]$, because $q(x)$ is irreducible in $F[x]$. Then at least one of the coefficients in $r(x)$ is not in F.

$$r(x) = (x - \theta_1)(x - \theta_2) \cdots (x - \theta_k)$$
$$= x^k - \sum_{i=1}^{k} \theta_i x^{k-1} + \sum_{i \neq j} \theta_i\theta_j x^{k-2} + \cdots$$
$$+ (-1)^m \sum_{\substack{\text{sum over products} \\ \text{of } m \text{ different } \theta\text{'s}}} \theta_i\theta_j \cdots \theta_m x^{k-m} + \cdots + (-1)^k \theta_1\theta_2 \cdots \theta_k.$$

Suppose the coefficient r_m of x^{k-m} is not in F. This coefficient is the sum of all products of m of the first k conjugates of θ. Since it is not in F, there is an automorphism σ_k in $G(E/F)$ that moves the coefficient: $\sigma_k r_m \neq r_m$. But this implies that σ_k moves one of the k conjugates of θ to a conjugate θ_{k+1}, since otherwise σ_k would merely permute the terms of r_m.

We conclude by finite induction that $q(x) = (x - \theta_1)(x - \theta_2) \cdots (x - \theta_d)s$, where s has degree zero and all θ's are in E. Then each irreducible $q(x)$ with a zero in E splits in E, proving that E is normal. ∎

The following theorem is our third necessary and sufficient condition for normality.

Theorem J–2. Let E be a finite extension of F, a field of characteristic zero. Then E is a normal extension if and only if E is the splitting field for a polynomial $f(x) \in F[x]$ of positive degree.

PROOF: From Theorem I–4, E equals $F(\psi)$, a simple algebraic extension of F with primitive element ψ. Let $p(x)$ be the monic irreducible polynomial in $F[x]$ having ψ as a zero, as in Theorem 7–1. If E is normal, then $p(x)$ splits in E. But $p(x)$ cannot split in a proper subfield of E, for a subfield containing the zero ψ must contain all of $E = F(\psi)$. Then E is the splitting field for $p(x)$ over F.

Conversely, let E be the splitting field for a polynomial $f(x) \in F[x]$. Then $E \cong F(\theta_1, \theta_2, \ldots, \theta_n)$, where the θ's are the zeros of $f(x)$ in E (Theorem 7–5). Let $p(x)$ be any irreducible polynomial in $F[x]$ having a zero ϕ in E. We want to prove that $p(x)$ splits in E, which by definition means that E is normal over F. If $\deg p(x) = 1$, then $p(x) = c(x - \phi)$, $c \in F$, and $p(x)$ splits in E. Suppose $\deg p(x) > 1$, so that $p(x)$ has another zero, say ψ, in some extension D of E. From Theorem I–1, $F(\phi) \cong F(\psi)$. Since $\phi \in E$, we have $E = E(\phi)$. $E = E(\phi)$ is a splitting field for $f(x)$ over $F(\phi)$, because (1) $f(x)$ splits in E by hypothesis and (2) an intermediate field in which $f(x)$ splits must contain ϕ and the θ's, hence all of E.

Also, $E(\psi)$ is a splitting field for $f(x)$ over $F(\psi)$, because (1) $f(x)$ splits in E, hence in $E(\psi)$, and (2) an intermediate field in which $f(x)$ splits must contain ψ and the θ's, hence all of $E(\psi)$.

Summing up, $f(x)$ has splitting fields E over $F(\phi)$ and $E(\psi)$ over $F(\psi)$ and $F(\phi) \cong F(\psi)$. Then by Theorem I–2, E and $E(\psi)$ are isomorphic. Then $[E:F] = [E(\psi):F]$, so that $[E(\psi):E] = 1$, and $\psi \in E$. Similarly, all the zeros of $p(x)$ lie in E, so that $p(x)$ splits in E. Then E is normal over F. ∎

> **Corollary.** Let D be a normal extension of F, a field of characteristic zero, and suppose $F \subseteq E \subseteq D$. Then D is normal over E.

PROOF: From Theorem J–2, D is the splitting field of a polynomial $f(x) \in F[x]$. The extension E of F has characteristic 0, and D is a finite extension of F, since it is a normal, and therefore a finite extension of E. The polynomial is also in $E[x]$, since each coefficient is in F, therefore also in E, and $f(x)$ splits in D and in no proper subfield. Then D is a splitting field for $f(x)$ over E and by Theorem J–2, D is normal over E. ∎

Exercise J–1. Refer to Exercise 8–3 concerning $Q(\psi)$, where $\psi = \sqrt{2} + fi$. Form $p(x) = (x - \psi_1)(x - \psi_2)(x - \psi_3)(x - \psi_4)$ and so find a polynomial in $Q[x]$ with a zero in D. Show that D is the splitting field for $p(x)$ over Q, hence that D is normal over Q.

> **Theorem 8–4** (The Fundamental Theorem of Galois Theory). Let D be a normal extension of F, a field of characteristic zero. Let E be an intermediate field, $F \subseteq E \subseteq D$. Then the following four statements hold:
>
> i. $E \leftrightarrow G(D/E)$ provides a one-to-one correspondence between all the intermediate fields E and all the subgroups of $G(D/F)$.
>
> ii. $[D:E] = |G(D/E)|$ and $[E:F] = [G(D/F):G(D/E)]$.

iii. $D \supseteq E_2 \supset E_1 \supseteq F$ if and only if $\hat{1} \subseteq G(D/E_2) \subset G(D/E_1) \subseteq G(D/F)$. Here "$\subset$" denotes proper inclusion, not equality.

iv. E is a normal extension of F if and only if $G(D/E)$ is a normal subgroup of $G(D/F)$, in which case

$$G(E/F) \cong G(D/F)/G(D/E).$$

PROOF: *i.* By Theorem J–2 and its corollary, we can sharpen the correspondences of Theorem 8–1. For normal extensions the two mappings of Theorem 8–1 are inverses of each other and so provide a one-to-one correspondence covering respectively all the intermediate fields and all the subgroups of $G(D/F)$. (See Figure 8–2.)

Let E be any intermediate field. Then there is a corresponding subgroup $G(D/E)$, the automorphisms in $G(D/F)$ that fix E. By the Corollary to Theorem J–2, D is normal over E, so by Theorem J–1, E is the fixed field of $G(D/E)$. Then every intermediate field corresponds to exactly one subgroup. Is every subgroup covered? Let H be any subgroup of $G(D/F)$. Let E be the fixed field of H. Then $H \subseteq G(D/E)$, for each automorphism in H fixes E. Since D is normal over F, it is normal over the intermediate field E, by the Corollary to Theorem J–2. Then by Theorem 8–3, $|G(D/E)| = [D:E]$. Let H have h automorphisms $\hat{1} = \sigma_1, \sigma_2, \ldots, \sigma_h$. Then if ψ is a primitive element in D over E, ψ is a zero of the polynomial $(x - \psi)(x - \sigma_2\psi) \cdots (x - \sigma_h\psi)$ of degree h. The coefficients of the polynomial are symmetric in the conjugates $\sigma_i\psi$, so they are fixed by H. Therefore, they are in E, the fixed field of H. Then ψ is a zero of a polynomial of degree h in $E[x]$, so $[D:E] \leq h$. But $h \leq [D:E] = |G(D/E)|$, so $h = |G(D/E)|$ and $H = G(D/E)$.

ii. From the Corollary to Theorem J–2, D is normal over E, so by Theorem 8–3 we have $[D:E] = |G(D/E)|$. From Theorem I–3, $[E:F] = [D:F]/[D:E] = |G(D/F)|/|G(D/E)| = [G(D/F):G(D/E)]$. (The index of a subgroup in a group is defined in Definition 3–8.)

iii. Notice that the reverse direction of containment of groups compared with the fields follows from the fact that fewer automorphisms fix more field elements. In the limiting cases, only the identity automorphism forming the 1-membered group $\langle\hat{1}\rangle$ fixes the whole field D, while every automorphism in $G(D/F)$ fixes the base field F.

Suppose $D \supseteq E_2 \supset E_1 \supseteq F$. We have $\langle\hat{1}\rangle = G(D/D) \subseteq G(D/E_2)$. Since $E_1 \subset E_2$, every automorphism in $G(D/E_2)$ also fixes E_1, $G(D/E_2) \subseteq G(D/E_1)$. But from property *ii*, $|G(D/E_2)| = [D:E_2] < [D:E_1] = |G(D/E_1)|$, so the groups must be different; $G(D/E_2) \subset G(D/E_1)$. Finally, each $G(D/E) \subseteq G(D/F)$.

Conversely, suppose $\langle\hat{1}\rangle \subseteq G(D/E_2) \subset G(D/E_1) \subseteq G(D/F)$. The fixed fields are D, E_2, E_1, and F, respectively, from which $D \supseteq E_2 \supset E_1 \supseteq F$, by property *i*.

iv. It is this property that enables us to reduce the solution of equations to individual steps. After each step we have an extension of the base field, still a normal extension so that we can again apply the theory of splitting fields.

Suppose E is a normal extension of F. Then $G(D/E)$ is a subgroup of $G(D/F)$, by Theorem 8–1. To prove it is a normal subgroup, we need to show that $\sigma G(D/E) = G(D/E)\sigma$ for each $\sigma \in G(D/F)$. That is, we must show that for each $\tau \in G(D/E)$, $\sigma^{-1}\tau\sigma \in G(D/E)$. Equivalently, we must show that $\sigma^{-1}\tau\sigma$ fixes E, since by the Corollary to Theorem J–2, D is normal over E, so that by Theorem J–1, E is the fixed field of $G(D/E)$. For $e \in E$ we must show that $\sigma^{-1}\tau\sigma(e) = e$, or $\tau[\sigma(e)] = \sigma(e)$. But we can prove that $\sigma(e)$ is in E: Since $\sigma \in G(D/F)$, e and σe are conjugates over F; that is, they are both zeros of an irreducible polynomial in $F[x]$ with a zero e in E. Since E is a normal extension, the polynomial splits in E, and σe is in E. Then since $\tau \in G(D/E)$, $\tau[\sigma(e)] = \sigma(e)$, proving that $G(D/E)$ is a normal subgroup of $G(D/F)$.

Conversely, suppose $G(D/E)$ is a normal subgroup of $G(D/F)$. Then the cosets $G(D/E)\sigma = \sigma G(D/E) = \sigma'$, for $\sigma \in G(D/F)$ form a quotient group $G(D/F)/G(D/E)$. For any $e \in E$, $[G(D/E)](\sigma(e)) = [G(D/E)\sigma](e) = \sigma' e = [\sigma G(D/E)](e) = \sigma[G(D/E)(e)] = \sigma(e)$, so $\sigma(e)$ is fixed by $G(D/E)$. Then $\sigma(e) \in E$, the fixed field of $G(D/E)$, according to Theorem J–1. Thus the automorphisms $\sigma' = \sigma G(D/E)$ fix E as a set and so provide automorphisms of E. The automorphisms σ' fix F, for if $f \in F$, then $\sigma' f = [\sigma G(D/E)](f) = \sigma(f) = f$. Then, as applied to E, $\sigma' \in G(E/F)$. In fact, as σ ranges over $G(D/F)$, the automorphisms σ' restricted to E include all the automorphisms of $G(E/F)$, for if $\rho \in G(E/F)$, $\tau \in G(D/E)$, then $\rho\tau^{-1} \in G(D/F)$ appears as a σ, for which $\sigma G(D/E) = \rho$. Because $G(D/E)$ is normal in $G(D/F)$, we also have $\sigma G(D/E) = G(D/E)\sigma = \rho$. Then $G(E/F) \cong G(D/F)/G(D/E)$.

Last, we must establish that if $G(E/F) \cong G(D/F)/G(D/E)$, then E is normal over F. The fixed field of $G(E/F)$ will be proved to be F, which by Theorem J–1 will suffice to prove E normal over F: Suppose e is fixed by all the automorphisms $\sigma G(D/E)$. Then $[\sigma G(D/E)](e) = \sigma(e) = e$, so that e is fixed by every $\sigma \in G(D/F)$. Then since D is normal over F, $e \in F$. Then F is the fixed field of $G(E/F)$, and E is a normal extension. ∎

THE CYCLIC GROUP OF
nth ROOTS OF UNITY

THEOREM 9-1. The nth roots of unity are distinct and form a cyclic group under multiplication.

PROOF: The nth roots of unity do form a group, for suppose r and s are nth roots of unity. Then the product rs is an nth root of unity, because $(rs)^n - 1 = r^n s^n - 1 = 1 \cdot 1 - 1 = 0$. The identity 1 and inverses $1/r$ are nth roots of unity, because $1^n - 1 = 0$ and $(1/r)^n - 1 = 1^n/r^n - 1 = 1/1 - 1 = 0$.

Write n in its prime-power factorization: $n = \prod_{i=1}^{s} d_i$, where $d_i = p_i^{a_i}$, p_i a prime, $a_i > 0$. For each i, $x^n - 1 = (x^{d_i} - 1)(x^{n-d_i} + x^{n-2d_i} + \cdots + x^{d_i} + 1)$. By Lemma 7-2 each of these polynomial factors in $Q[x]$ has no more zeros than its degree, so d_i of the nth roots of unity in $D \supseteq Q$ are also d_ith roots of unity, while $n - d_i$ of them fail to satisfy $x^{d_i} = 1$.

Factor $x^{d_i} - 1$ as

$$x^{d_i} - 1 = x^{p_i^{a_i}} - 1 = (x^{p_i^{a_i-1}} - 1)(x^{p_i^{a_i} - p_i^{a_i-1}} + \cdots + x^{p_i^{a_i-1}} + 1).$$

From this we can conclude that of the d_ith roots of unity $p_i^{a_i-1}$ are $p_i^{a_i-1}$st roots of unity, but the rest have order $d_i = p_i^{a_i}$ (recall Theorem 3-9). Let u_i have order d_i.

We can prove there are nth roots of unity having order n by exhibiting one, namely $r = \prod_{i=1}^{s} u_i$. Since this product is a member of the group, it does satisfy $x^n - 1$, so its order is a divisor of n. We show that it cannot be a proper divisor of n, for let $e = n/p$, where p is one of the prime divisors of n. Then

$$r^e = \left(\prod_{i=1}^{s} u_i\right)^e = \prod_{i=1}^{s} u_i^{n/p} = u^{n/p},$$

where $u = u_i$ for $p = p_i$, since if $p_i \neq p$, $d_i \mid n/p$. Since $u^{n/p} \neq 1$, the order of r is n.

Then the n distinct powers of r,

$$r, r^2, \ldots, r^{n-1}, r^n = 1,$$

constitute the nth roots of unity. ∎

PROOF THAT A SOLVABLE EQUATION HAS A SOLVABLE GALOIS GROUP

> **Theorem L-1.** Let $f(x) \in F[x]$, where F is a field of characteristic zero. Let $f(x) = 0$ be solvable by radicals. Then there is a root tower over F for a normal extension K_s that contains all the roots of $f(x) = 0$.

PROOF: The proof is by construction; we show how to construct on the basis of the given root tower (Definition 10–2) a root tower

$$F = K_0 \subseteq \cdots \subseteq K_1 \subseteq \cdots \subseteq K_{s-1} \subseteq \cdots \subseteq K_s,$$

with K_s a normal extension of F and all the roots of $f(x) = 0$ in K_s. Define K_0 equal to F. Define the polynomial

$$g_0(x) = x^{n_0} - f_0,$$

with n_0, f_0 from the given root tower. In the given root tower, $F_1 = F(w_1)$, where w_1 is a zero of $g_0(x)$. Adjoin any other zeros of $g_0(x)$ one by one to obtain K_1, the splitting field for $g_0(x)$ over F and a root tower for K_1 over F. In general, then the root tower from K_0 to K_1 contains several field extensions, say, $K_0 \subseteq K_{01} \subseteq \cdots \subseteq K_{0t} = K_1$. Notice that if F happens to contain the n_0th roots of unity, then the splitting field of $g_0(x)$ is just $F(w_1) = F_1$ (Theorem 9–3). In any case K_1 is a normal extension of F, because it is the splitting field for $g_0(x) \in F[x]$ (Theorem J–2), and $F_1 \subseteq K_1$.

Now suppose we have constructed K_i, normal over F, with a root tower over F, and containing F_i. To define K_{i+1} so that it is normal over F, we form

$$g_i(x) = (x^{n_i} - \sigma_1 f_i)(x^{n_i} - \sigma_2 f_i) \cdots (x^{n_i} - \sigma_m f_i),$$

with n_i, f_i from the given root tower, where $\sigma_1 = \hat{1}, \sigma_2, \ldots, \sigma_m$ are the automorphisms making up the Galois group $G(K_i/F)$. Adjoin the zeros of $g_i(x)$ to K_i in this way: First form the splitting field L_{i1} for $x^{n_i} - \sigma_1 f_i$ over K_i, then the splitting field L_{i2} for $x^{n_i} - \sigma_2 f_i$ over L_{i1}, and so on, with $L_{im} = K_{i+1}$ the splitting field for $x^{n_i} - \sigma_m f_i$ over $L_{i,m-1}$. The successive field extensions L_{ij} form a root tower from K_i to the splitting field K_{i+1} for $g_i(x)$ over K_i (Theorem 7–5). From Theorem J–1, the fixed field of $G(K_i/F)$ is F since K_i is normal over F. Then since all the coefficients in $g_i(x)$ are fixed by the σ's, they must lie in F. Consider the product polynomial

$$P = g_0(x)g_1(x) \cdots g_i(x).$$

Its coefficients lie in F, and its splitting field is K_{i+1}, which is F with all the zeros of P adjoined (Theorem 7–5). Then by Theorem J–2, as the splitting field for P over F, K_{i+1} is a normal extension of F. We have $F_{i+1} \subseteq K_{i+1}$, for $F_i \subseteq K_i \subseteq K_{i+1}$, and w_{i+1} as a zero of $x^{n_i} - \sigma_1 f_i$, $\sigma_1 = \hat{1}$, has been adjoined to construct K_{i+1}.

Since all the roots of $f(x) = 0$ lie in F_s and by this construction $F_s \subseteq K_s$, K_s satisfies the requirements of the theorem. ∎

> **Theorem L–2.** Continuing the notation of Theorem L–1, let $n = [n_0, n_1, \ldots, n_{s-1}]$, the least common multiple of the n's, and let u be a primitive nth root of unity. Define $E_0 = F(u)$. Let the root tower for K_s over F be relabeled to show each splitting field L_{ij} for $x^{n_i} - \sigma_j f_i$ over $L_{i,j-1}$, $1 \leq j \leq m$, in the construction of Theorem L–1:
>
> $$F = L_0 \subseteq L_1 \subseteq \cdots \subseteq L_t = K_s.$$
>
> If $L_{j+1} = L_j(v_{j+1})$, then define $E_{j+1} = E_j(v_{j+1})$. Then $G(E_t/E_0)$ is solvable.

PROOF: By Theorem 9–4, K_s contains all the n_ith roots of unity for $i = 0, 1, \ldots, s - 1$, because $x^{n_i} - f_i$ splits in K_s. Consequently, $u \in K_s$, $E_t = K_s$ since K_s by Theorem 9–4 contains u, and the extensions E_j form a root tower for E_t over E_0. $E_t = K_s$ is normal over F by Theorem L–1 and, hence, normal over each field E_j by the Corollary to Theorem J–2. Let $G_j = G(E_t/E_j)$, and consider the chain

$$G(E_t/E_0) = G_0 \supseteq G_1 \supseteq \cdots \supseteq G_t = \langle 1 \rangle.$$

Since L_{j+1} is a finite extension of L_j, it is a simple extension with some primitive element v_{j+1}. E_j contains all the n_ith roots of unity and E_{j+1} is by the construction of L_{j+1} a splitting field for $x^{n_i} - \sigma_j f$ over L_j, so by Theorem 9–3, $E_{j+1} = E_j(v_{j+1})$, and $G(E_{j+1}/E_j)$ is cyclic. Then by the Fundamental Theorem of Galois Theory (Theorem 8–6iv), G_{j+1} is normal in G_j (or $G_j \triangleright G_{j+1}$) and $G(E_{j+1}/E_j) = G_j/G_{j+1}$, for each $j = 0, 1, \ldots, t - 1$. Then the respective quotient groups are cyclic, hence certainly abelian. If we omit duplicate listings of equal groups, we have for $G_0 = G(E_t/E_0)$ the chain required by Definition 10–3, proving that G_0 is solvable. ∎

We are now very close to proving that if $f(x) = 0$ is solvable by radicals, then its Galois group is a solvable group. Its Galois group is $G(D/F)$, where D is a splitting field for $f(x)$ over its coefficient field F. We have proved instead that $G(E_t/E_0)$ is solvable (Theorem L–2), where $E_t \supseteq D$, both normal over F, and $E_0 = F(u)$, u a primitive nth root of unity.

We prove next a result about solvable groups that will help us establish the solvability of $G(D/F)$.

Lemma L–I.   Let G be a finite group with a normal subgroup S. If G is solvable, then G/S is solvable.

PROOF: From Definition 10–3, there is a chain of subgroups for G

$$G = G_0 \triangleright G_1 \triangleright \cdots \triangleright G_t = \langle 1 \rangle,$$

such that G_i/G_{i+1} is abelian. Let g and h be elements of G_i. In coset form the statement that G_{i+1} is normal in G_i is written

$$gG_{i+1} = G_{i+1}g. \tag{1}$$

Since the coset elements of G_i/G_{i+1} are gG_{i+1}, hG_{i+1}, and so forth, the statement that the quotient group is abelian has the coset form

$$(gG_{i+1})(hG_{i+1}) = (hG_{i+1})(gG_{i+1}). \tag{2}$$

Since S is given normal in G, we have

$$gS = Sg. \tag{3}$$

We use the commutativity relations of (1), (2), and (3) to prove that there is a chain of subgroups leading from G to S.

Let $G_iS = \{gs \mid g \in G_i, s \in S\}$. Then G_iS is a subgroup of G by Theorem 3–6, because for any $g, h \in G_i$ and $s, t \in S$, $gs \cdot ht = g(sh)t$ by associativity. From (3) there is an $r \in S$ for which $sh = hr$. Then $gs \cdot ht = g(hr)t = (gh)(rt) = g's'$ for some $g' \in G$, $s' \in S$. The inverse of $gs = s^{-1}g^{-1} = g^{-1}r'$, by (3), which is in G_iS.

Since each $g_{i+1}S \in G_iS$, where $g_{i+1} \in G_{i+1}$, because $G_i \supset G_{i+1}$, we have $G_iS \supset G_{i+1}S$. Form the chain

$$G = G_0S \supset G_1S \supset \cdots \supset G_tS = \langle 1 \rangle S = S.$$

Since $G_{i+1}S$ is a subgroup of G all of whose members lie in G_iS, $G_{i+1}S$ is a subgroup of G_iS. In fact, it is normal in G_iS, for $(gS)(G_{i+1}S) = SgG_{i+1}S$ (by associativity and (3)) $= SG_{i+1}gS$ (by (1)) $= (G_{i+1}S)(gS)$ (by (3)). Each quotient group $G_iS/G_{i+1}S$ is abelian, because the product of two coset members of the quotient group is $(gSG_{i+1}S)(hSG_{i+1}S) = SgG_{i+1}hG_{i+1}SSS$ (by associativity and (3)) $= S(hG_{i+1})(gG_{i+1})SSS$ (by (2)) $= (hSG_{i+1}S)(gSG_{i+1}S)$ (by (3)).

Now we form the quotient group G_iS/S for each i in the chain. Since the elements of G_iS/S have the form $gsS = gS$, we have the inclusions, normal subgroups, and abelian quotient groups necessary for a solvability chain for G/S:

$$G/S = G_0S/S \rhd G_1S/S \rhd \cdots \rhd G_tS/S = S/S = \langle 1 \rangle. \quad \blacksquare$$

Theorem L–3. Let $f(x) = 0$ be solvable by radicals, where $f(x) \in F[x]$ and F is a field of characteristic zero. Then the Galois group of $f(x)$ is solvable.

PROOF: Our first objective is to prove that $G(E_t/F)$ is solvable. By Theorem 9–2, $G(E_0/F)$ is abelian. (We continue the notation used previously. $E_0 = F(u)$, where u is a primitive nth root of unity.) Since E_0 is normal over F, Theorem 8–4iv gives us $G(E_0/F) \cong G(E_t/F)/G(E_t/E_0)$. If $E_0 \neq F$, then

$$G(E_t/F) \rhd G(E_t/E_0) = G_0 \rhd \cdots \rhd G_{s'} = \langle 1 \rangle,$$

where the chain for G_0 is the one found in Theorem L–2. $G(E_t/F)/G_0$ is abelian because it is isomorphic to $G(E_0/F)$, which is abelian. Each of the later quotients G_j/G_{j+1} was proved abelian in Theorem L–2. Then we have the chain of groups required by Definition 10–3 to prove that $G(E_t/F)$ is solvable.

Now we have $F \subseteq D \subseteq E_t$, E_t is normal over F, and D, as the splitting field for $f(x)$, is normal over F, so we can apply Theorem 8–4iv to conclude that

$$G(D/F) \cong G(E_t/F)/G(E_t/D).$$

Then by Lemma L–1 we can conclude from the solvability of $G(E_t/F)$ that $G(D/F)$ is solvable. $\quad \blacksquare$

IMPOSSIBLE CONSTRUCTIONS

It is possible to prove that in general an angle cannot be divided into equal thirds by the kind of ruler-and-compass construction used in plane geometry. This negative result is similar to Abel's Theorem and uses some of the same theory.

Suppose you have a straightedge, which is like a ruler that is not marked off in units, and a compass. Then as in plane geometry you can bisect angles, construct perpendiculars, draw an angle congruent to a given angle, and so forth. Since each angle in an equilateral triangle is 60°, you can construct a 60° angle as in Figure M–1, by drawing C_1 and then without changing the compass setting drawing circle C_2 with its center on C_1.

Then you can construct a 30° angle by bisecting the 60° angle, as in Figure M–2.

However, the seemingly simple task of constructing a 20° angle is impossible. When we say that a 20° angle is not **constructible**, we mean it in a sense just like that of Abel's Theorem; that is, we mean it cannot be constructed in a certain context. In this case, we mean that a finite construction process involving straightedge and compass alone can never produce a 20° angle.

How can this negative result be proved? First we show what we can expect in case a length *is* constructible.

Theorem M–1. If a length is constructible it lies in some extension field F_c, where $F_0 = Q$, the rationals, and $F_{i+1} = F_i(w_{i+1})$, where $w_{i+1}^2 = f_i \in F_i$, $i = 1, 2, \ldots, c - 1$. (Reread Definition 10–2 and compare.)

PROOF: First, every length in $F_0 = Q$ is constructible, choose some distance (in the form of some compass setting) as a fundamental unit 1, mark it off n times against the straightedge to construct any positive integer and, hence, construct each rational a/b as in Figure M–3.

Let F_i stand not only for the field extension, or $F_0 = Q$ for $i = 0$, but also for all the points (x, y) on a Cartesian graph with $x \in F_i$ and $y \in F_i$. As we demonstrated in Figure M–3, each point of F_0 is constructible. Let a "line in F_i" mean a line connecting two points that lie in F_i; let a "circle in F_i" mean a circle with center in F_i and radius r in F_i.

As in Figure M–4 we use the fact that respective sides of similar triangles are proportional to obtain an equation for a line in F_i connecting (x_1, y_1) and (x_2, y_2), where all four of these coordinates lie in F_i. If $x_1 = x_2$ the equation is just $x = x_1$.

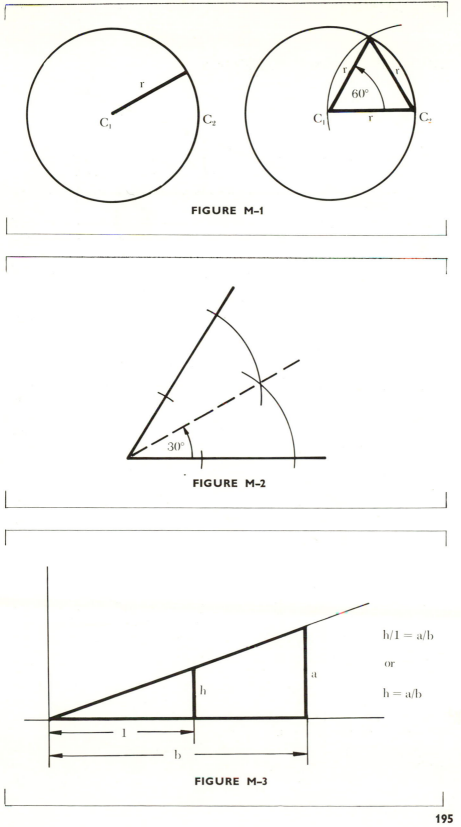

FIGURE M–1

FIGURE M–2

$$h/1 = a/b$$

or

$$h = a/b$$

FIGURE M–3

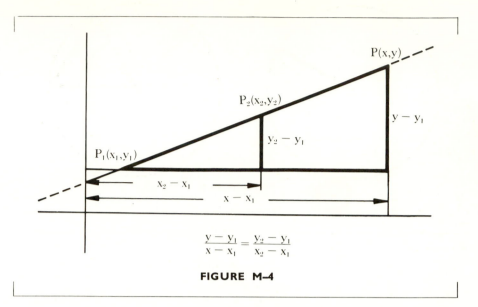

$$\frac{y - y_1}{x - x_1} = \frac{y_2 - y_1}{x_2 - x_1}$$

FIGURE M–4

As in Figure M–5 we use the Pythagorean Theorem to obtain an equation for a circle in F_i.

Exercise M–1.　Construct the points $P_1(0, \frac{5}{2})$ and $P_2(3, 1)$, both in Q.

Exercise M–2.　Find an equation for the line in Q through the points in Exercise M–1.

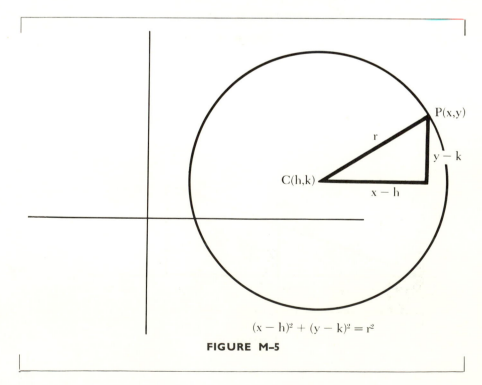

$$(x - h)^2 + (y - k)^2 = r^2$$

FIGURE M–5

Exercise M–3. Find an equation for the line in Q through $(-1, 0)$ and $(\frac{3}{2}, \frac{3}{2})$.

Exercise M–4. Show that the lines in Exercises M–2 and M–3 intersect at $(\frac{19}{11}, \frac{18}{11})$, a point in Q.

The intersection of two lines in F_i is a point in F_i. An intersection of a circle in F_i with a line in F_i or with another circle in F_i is a point in F_{i+1}: To prove the first statement, notice that each linear equation yields an expression for y in terms of x, all coefficients of which are in F_i. Equating two of these expressions to eliminate y, we solve for x. This solving process involves only rational operations, but F_i is closed under its rational operations, so the results must again lie in F_i.

To prove that an intersection of a line in F_i and a circle in F_i is a point in F_{i+1}, notice that the linear equation yields an expression for y in terms of x, all coefficients in F_i. Substitute this expression for y in the equation of the circle. The result is a quadratic equation in x, all of whose coefficients are in F_i. From the theory of quadratic equations we have been studying, then, the intersection lies in F_{i+1}. (Can it ever lie in F_i? In that case is it correct to say it lies in F_{i+1}?)

To prove that an intersection of two circles in F_i is a point in F_{i+1}, subtract one equation from the other to eliminate both x^2 and y^2. The result is a linear equation with coefficients in F_i, which is then solved simultaneously with either of the circle equations as in the previous paragraph. ∎

Theorem M–2. The length $\cos 20°$ is not constructible. An angle of $20°$ is not constructible.

PROOF: We outline the proof here, reserving details for exercises.
From the trigonometric identity

$$\cos 3A = 4 \cos^3 A - 3 \cos A,$$

applied to $A = 20°$, we have

$$\tfrac{1}{2} = \cos 60° = 4 \cos^3 20° - 3 \cos 20°,$$

letting c stand for $\cos 20°$,

$$8c^3 - 6c - 1 = 0.$$

Then $c = \cos 20°$ is a zero of an irreducible equation of degree 3, so that $[F(c):F_0] = 3$.

However, the degree of each extension F_i in Theorem M–1 is a power of 2, and 3 fails to divide any power of 2. Then c is not constructible.

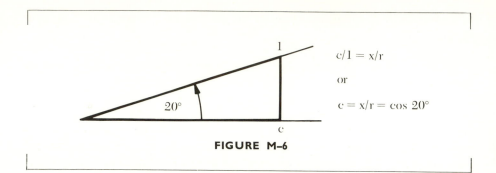

FIGURE M–6

But if it were possible to construct an angle of 20°, then it would be possible to construct its cosine as in Figure M–6. Then the angle is also not constructible.

Exercise M–5. Use the trigonometric identities

$$\cos(A + B) = \cos A \cos B - \sin A \sin B$$
$$\sin(A + B) = \sin A \cos B + \cos A \sin B$$

several times to prove the identity

$$\cos 3A = 4 \cos^3 A - 3 \cos A.$$

Exercise M–6. Use the Corollary to Theorem 6–6 to show that $8c^3 - 6c - 1$ is irreducible over the rationals.

Exercise M–7. What theorem in this book justifies the assertion $[F(c):F_0] = 3$?

Exercise M–8. Use Theorem 7–7 to justify the assertion that $[F_i:F_0]$ is a power of 2.

Exercise M–9. Use Theorem 7–7 in the case $F_0 \subset F_0(c) \subseteq F_i$ to show that if c lay in some F_i, then $3 = [F_0(c):F_0]$ would have to divide $[F_i:F_0]$. ∎

There are several other famous construction problems that have been proved impossible by methods similar to this. A good description appears on pages 187–190, I. N. Herstein, *Topics in Algebra*, Blaisdell, 1964.

INDEX

Page on which definition appears is distinguished by **boldface** type.

SYMBOLS